# Machining Projects

### David O. Averyt

Maxwell High School of Technology
Lawrenceville, Georgia

Publisher
**The Goodheart-Willcox Company, Inc.**
Tinley Park, Illinois
www.g-w.com

Copyright © 2007
by
The Goodheart-Willcox Company, Inc.

All rights reserved. No part of this work may be reproduced, stored, or transmitted in any form or by any electronic or mechanical means, including information storage and retrieval systems, without the prior written permission of The Goodheart-Willcox Company, Inc.

Manufactured in the United States of America.

ISBN-13: 978-1-59070-779-1
ISBN-10: 1-59070-779-6

3 4 5 6 7 8 9 – 07 – 18 17 16 15

**The Goodheart-Willcox Company, Inc. Brand Disclaimer:** Brand names, company names, and illustrations for products and services included in this text are provided for educational purposes only and do not represent or imply endorsement or recommendation by the author or the publisher.

**The Goodheart-Willcox Company, Inc. Safety Notice:** The reader is expressly advised to carefully read, understand, and apply all safety precautions and warnings described in this book or that might also be indicated in undertaking the activities and exercises described herein to minimize risk of personal injury or injury to others. Common sense and good judgment should also be exercised and applied to help avoid all potential hazards. The reader should always refer to the appropriate manufacturer's technical information, directions, and recommendations; then proceed with care to follow specific equipment operating instructions. The reader should understand these notices and cautions are not exhaustive.

The publisher makes no warranty or representation whatsoever, either expressed or implied, including but not limited to equipment, procedures, and applications described or referred to herein, their quality, performance, merchantability, or fitness for a particular purpose. The publisher assumes no responsibility for any changes, errors, or omissions in this book. The publisher specifically disclaims any liability whatsoever, including any direct, indirect, incidental, consequential, special, or exemplary damages resulting, in whole or in part, from the reader's use or reliance upon the information, instructions, procedures, warnings, cautions, applications, or other matter contained in this book. The publisher assumes no responsibility for the activities of the reader.

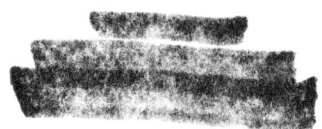

*Machining Projects*

# Table of Contents

**Introduction** .................................................. 5
    Using This Book. ......................................... 6
    To the Instructor ......................................... 9
    To the Student Machinist. ............................... 10
    Shop Safety. ............................................ 10

**Reference Material** ........................................... 15

**Section 1—Bench Work** ...................................... 25
    1.1    Drill Gage. ...................................... 26
    1.2    Drill Drift. ...................................... 28
    1.3    Drill Index Block. ............................... 30
    1.4    Drill Index Block. ............................... 33
    1.5    Tap Index Block. ................................ 36
    1.6    Tapping Plate .................................. 39

**Section 2—Lathe Competency** ............................... 41
    2.1    Faced Bar ..................................... 42
    2.2    Faced Bar—Drilled and Hand Tapped ............ 44
    2.3    Step Bar ...................................... 46
    2.4.1  Threaded Bar .................................. 48
    2.4.2  Threaded Nut ................................. 50
    2.5.1  Turned Bar—Undercut and Chased Threads ...... 52
    2.5.2  Threaded Nut—Undercut and Chased Threads ... 54
    2.6    Knurled Bar ................................... 56
    2.7    Go, No-Go Bar ................................ 58
    2.8    Bored Bar. .................................... 60
    2.9.1  Tapered Bar ................................... 62
    2.9.2  Tapered Sleeve ................................ 64

## Section 3—Vertical Mill Competency .................................... 66

| | | |
|---|---|---|
| 3.1 | Faced Bar | 67 |
| 3.2 | Drilled Plate | 69 |
| 3.3 | Drilled Plate | 71 |
| 3.4 | Step Plate | 73 |
| 3.5 | Slotted Block or Bar | 75 |
| 3.6 | Multi-Operations Bar | 77 |
| 3.7 | Angle Bar | 79 |
| 3.8 | Round Bar | 81 |
| 3.9 | Mill Block—Drilled and Bored Holes | 83 |
| 3.10 | Fluted Round Bar | 85 |

## Section 4—Projects .................................................... 87

| | | |
|---|---|---|
| 4.1 | Barstock Spindle Stop Adapter for Lathe | 88 |
| 4.2 | Ball Peen Hammer | 90 |
| 4.3 | Center Drill Chuck | 96 |
| 4.4 | Deburring Tool | 98 |
| 4.5 | Gravity Center Punch | 109 |
| 4.6 | Fly Cutter | 114 |
| 4.7 | Lathe Puzzle | 116 |
| 4.8 | Mill Cutter Arbor | 122 |
| 4.9 | Model Civil War Cannon | 124 |
| 4.10 | R-8 Tool Holder | 130 |
| 4.11 | Machinist Screw Jack | 132 |
| 4.12 | Meat Tenderizing Hammer | 139 |
| 4.13 | Paper Punch | 145 |
| 4.14 | Tap Wrench | 154 |
| 4.15 | Step Block | 158 |
| 4.16 | Parallels | 160 |
| 4.17 | Bolt Welding Jig | 164 |
| 4.18 | Air Engine | 168 |

# Introduction

The process that led to the publication of this book began during the author's first experience at the SkillsUSA National Competitions. While his student participated in the precision machining competition, he circulated through the vendor area in hopes of locating a machining project book that had something new he could incorporate into his program. After making the rounds of the publisher booths, he was told that there was not anything on the market that consisted primarily of projects. There are many excellent textbooks, workbooks, and lab manuals that address the concepts, knowledge, and skills that collectively make up the field of machine tool operation, but a book containing projects specifically designed for a machining technology program was not available. When he expressed frustration at not being able to find such a book, the publishing representative with whom he was talking asked, "Why don't you write one?" The author's first reaction was that an undertaking of that magnitude was outside the realm of possibility. Then, after further consideration, he decided with his many years of experience teaching in a machining program, this was a task that he could take on.

At the onset of the development process, he had intended that it be simply a collection of machining projects that instructors could use as a resource when planning their instruction. As he began to compile project plans and finalize the drawings of projects of his own design, he came to realize that here was a greater opportunity. This was to put something together that not only provided a fresh collection of machining projects, but could also address some of the instructional problems that are encountered with students at both the secondary and postsecondary level. These problems include:

- The difficulty many student machinists have in visualizing or understanding what to do next.
- Not having a complete knowledge or grasp of the details of a particular machining operation, concept, or procedure.
- The tendency to depend on the instructor for answers, rather than learning to use the textbook as a reference, reviewing the applicable sections, and then thinking through the problem and finding a solution.

With these common problems in mind, the author set out to design a sequence of competency development pieces for each of the major machine tool areas. These competency development sections begin with the basic operations relating to that particular piece of equipment, and then progress through the major operations it can perform. In addition to the competency sections, the second half of this book is a collection of machining projects from which the instructor and student can choose assigned lab projects. Each of these projects is accompanied by an order of operations to aid individuals in its completion. While many of the projects in this section are time-honored favorites, a significant number of them are of the author's design and are being published for the first time.

This book, in both content and format, is based on three key concepts: functionality, flexibility, and friendliness. Above all else, *this book is meant to be used*. The author has addressed the concept of *functionality* in a number of ways. First, each competency piece, as well as each project, is accompanied by a suggested order of operations. Operations in the competency sections are referenced to the appropriate section in the Goodheart-Willcox textbook, **Machining Fundamentals**. This textbook is where an explanation of that operation may be found. The operations are also referenced to the National Institute for Metalworking Skills (NIMS) *Duties and Standards for Machining Skills*. The projects in the *Projects* section do not include these references. The student can refer back to the appropriate competency should a problem arise. In addition, a section of the operations page has been lined and set aside for notes. With the order of operations and the references readily available, the student has the opportunity and responsibility to review the appropriate sections of the text when necessary, rather than seek a quick answer from the instructor. This provides student machinists the opportunity to practice using reference materials when seeking the solution to a problem.

In the best of all possible worlds, everyone would have access to a lab equipped with a full complement of new machines. In reality, however, many do all that they can with the resources available to them. For this reason, the second key concept that has guided the development of this book is *flexibility*. The format of this book is designed to allow an instructor to assign specific operations, sequences of operations, or projects according to available equipment and resources. An instructor can also assign particular competency pieces as remediation should it be necessary. In the *Projects* section, the number and selection of projects gives the instructor the ability to tailor the project assignments to a wide variety of student interests and abilities. With discreet competency sections, the instructor can also form teams assigned to a particular machine section to better distribute students around the lab and lessen the possibility of having to wait excessive periods of time for the opportunity to complete their assignments.

The third concept that underlies the format and content of this book is *friendliness*. At first glance, it might seem odd that the idea of friendliness should be a consideration in a book of this type. However, the author developed this book as something he would like to have to aid him in teaching and reinforcing the competencies necessary to survive and prosper in the machining industry. Whenever possible, he tried to make this book as easy to use as he could. He has done this for the student by designing a format in which the competency-building process occurs in small increments. The drawings, the orders of operations, and where to find information on those operations are immediately available. The competency pieces have been kept small to conserve materials and relatively simple to keep the student machinist from being overwhelmed by drawings that have a multitude of notes, lines, and dimensions. Although more than one operation could be performed on a competency piece, the author designed them as individual pieces. This is done so students will have to perform the fundamental operations that are done to every workpiece a number of times. Common operations—such as cutting to length plus facing allowance, facing, and center drilling—will be performed repetitively and will hopefully be internalized.

In the *Projects* section, the author also has tried to find or design projects that appeal to a wide spectrum of student interests. He believes that many of the things that make this book student-friendly also make it instructor-friendly as well. This is a single volume containing sequential series of competency-developing lab experiences that are referenced to the **Machining Fundamentals** textbook, provides direction in the order of operations, and is material conservative. All this and a wide variety of standard and new machining projects will make machining a tremendous learning experience for students and instructors alike.

## Using This Book

This book is a resource designed for the instructor and the student. It is meant to be a source of information and inspiration as well. This book is divided into the four sections: *Bench Work*, *Lathe Competency*, *Vertical Mill Competency*, and *Projects*.

### Bench Work

It is common knowledge that CNC (computer numerical control) machine tools are becoming the norm in the machining industry. However, every machinist will, at some time or another, be called upon to perform tasks that require the use of various hand tools. Some of these tasks could include such operations as layout, hand filing and finishing, threading and tapping of bolts and holes, hand polishing, and assembly and disassembly of machines and equipment. These tasks, and the skills and knowledge required to perform them, are known collectively as **bench work**. Manual skill in these operations can only be developed through practice. For this reason, this book contains several bench work projects. In the case of each of these projects, the band saw will be used to cut the workpiece to the rough shape and size, and the drill press used for drilling any required holes. Also, a variety of hand tools will be used to bring the project pieces to final size, shape, and degree of polish.

*Machining Projects*

## Bench Work Task List
1. Identify the hand tools commonly used in metalworking.
2. Demonstrate the proper use of hand tools.
3. Demonstrate the ability to perform basic shop math calculations.
4. Demonstrate the ability to read, interpret, and work from basic blueprints.
5. Demonstrate proper layout and marking techniques.
6. Demonstrate the correct use of the drill press, including proper safety precautions, as well as commonly used accessories.
7. Demonstrate knowledge of abrasive materials and the correct techniques for hand polishing metal surfaces.
8. Demonstrate the ability to measure accurately to 1/64″.
9. Demonstrate the correct use of the horizontal and vertical band saws, including proper safety precautions and commonly used accessories.
10. Demonstrate the correct use of the pedestal and belt grinders.

## Lathe and Vertical Mill Competency

The *Lathe Competency* section and the *Vertical Mill Competency* section are designed to help the student machinist become competent in all major operations commonly performed on the lathe and the vertical mill. The competency pieces are designed to provide the opportunity to practice and master each of the major operations routinely performed on the manual lathe and the vertical milling machine. The lathe sequence of operations are presented from the most basic of lathe operations, such as facing a workpiece to a specified length, to increasingly more complex operations, such as turning a specified length and diameter and chasing internal and external threads. The vertical milling sequence of operations are presented from the most basic of mill operations, such as facing a workpiece to a specified length, to increasingly more complex operations, such as stepping off dimensions, milling and drilling, and boring. For each competency workpiece, a print (drawing) of the piece and an order of operations are provided. With each operation there is a reference number that indicates where in the ***Machining Fundamentals*** textbook information may be found about the particular operation about to be performed. There are also areas on the pages so notes can be taken from the textbook or from the instructor's lectures.

## Lathe Work Task List
1. Demonstrate proper safety practices and precautions in lathe operations.
2. Measure with and read a micrometer (outside, inside, depth).
3. Measure with and read vernier and dial calipers.
4. Check alignments and dimensions with a dial indicator.
5. Properly grind single-point cutting tools for roughing, finishing, threading, grooving, and cut-off operations.
6. Chuck workpiece in 3-jaw universal, 4-jaw independent, and collet chucks.
7. Calculate and set proper speeds and feeds.
8. Properly set up cutting tools for facing and turning operations.
9. Prepare and set up workpieces for turning between centers.
10. True lathe tailstock and workpieces using dial indicator.
11. Demonstrate proper technique for filing workpiece on the lathe.
12. Calculate, set up, and turn internal and external tapers on the lathe using compound rest, taper attachment, and tailstock offset.
13. Cut internal and external threads using taps and dies.
14. Demonstrate knurling setup and operations.
15. Set up and center drill, bore, counterbore, and countersink holes on a lathe.
16. Perform reaming operations on a lathe.
17. Set up and chase internal and external threads on a lathe.
18. Demonstrate correct use of thread measuring tools.
19. Perform cut-off operations on a lathe.
20. Maintain the lathe for proper operation.

## Milling Machine Task List

1. Demonstrate proper safety practices and precautions for milling operations.
2. Align head of milling machine using a dial indicator.
3. Align vise on mill table using a dial indicator.
4. Align workpiece with a dial indicator.
5. Select and mount proper workholding devices.
6. Properly mount workpieces in vise or on mill table.
7. Select and install proper milling attachments.
8. Identify, select, and install proper milling cutters.
9. Calculate and set milling machines for proper speeds, feeds, and depth of cut.
10. Square stock to given specifications.
11. Machine surfaces with proper milling cutters.
12. Machine slots and steps using an end mill.
13. Machine flat surfaces with a fly cutter.
14. Perform drilling and reaming operations.
15. Bore a hole to specifications using a boring head.
16. Machine a workpiece to a specified angle and height.
17. Read, interpret, and use advanced blueprints.
18. Maintain milling machines for proper operation.

## The NIMS Standards

The competency pieces in this book have been referenced to the corresponding section in the National Institute for Metalworking Skills's (NIMS) *Duties and Standards for Machining Skills–Level One*. The NIMS Standards, which were originally developed by the National Tooling and Machining Association (NTMA), include seven categories of occupational duties or skills that make up the level one certification. These seven categories are:

**Section 1. Job Planning and Management**
    1.1 Job Process Planning

**Section 2. Job Execution**
    2.1 Manual Operations: Benchwork
    2.2 Manual Operations: Layout
    2.3 Turning Operations: Between Centers Turning
    2.4 Turning Operations: Chucking
    2.5 Milling: Square Up a Block
    2.6 Vertical Milling
    2.7a Grinding Wheel Safety
    2.7b Surface Grinding
    2.8 Drill Press
    2.9 CNC Programming

**Section 3. Quality Control and Inspection**
    3.1 Part Inspection
    3.2 Process Control

**Section 4. Process Adjustment and Control**
    4.1 Process Adjustment—Single Part Production
    4.2 Participation in Process Improvement

**Section 5. General Maintenance**
    5.1 General Housekeeping and Maintenance
    5.2 Preventive Maintenance
    5.3 Tooling Maintenance

Section 6. **Industrial Safety and Environmental Protection**
        6.1   Machine Operations and Material Handling
        6.2   Hazardous Material Handling and Disposal
Section 7. **Career Management and Employment Relations**
        7.1   Career Planning
        7.2   Job Applications and Interviewing
        7.3   Teamwork and Interpersonal Relations
        7.4   Organizational Structures and Work Relations
        7.5   Employment Relations

Based on the focus of this book, the references are limited to *Section 1—Job Planning and Management* and *Section 2—Job Execution*.

The NIMS *Duties and Standards for Machining Skills* will prove to be a valuable and useful tool for the enhancement of the knowledge and skill of all who are involved in machining. If you wish to know more about NIMS and the resources available they offer, visit their Web site at *www.nims-skills.org*.

## Projects

The *Projects* section contains projects that will appeal to a wide range of student interests. These machining projects are designed to provide a student the opportunity to apply the knowledge attained and the skills developed in the *Bench Work, Lathe Competency*, and *Vertical Mill Competency* sections. Many of these projects are tools that will enhance a machinist's work, while others are more decorative in nature. Whatever their primary form or function, if they are to be done well, each will require skill, knowledge, and attention to detail. This is the hallmark of craftsmanship.

While some of these projects can be completed using just one type of machine, most will require the use of several different ones. The order of operations that accompany each project will help with the sequence of the machining operations. All of the operations required to complete these projects have been covered in the competency sections. Although there are no specific references, referring back to the competency sections for review of a particular operation is recommended.

# To the Instructor

This book is intended to offer the instructor the flexibility to assign tasks and projects that address students' specific needs and the equipment capabilities of individual labs or instructional situations. With that taken into consideration, the following suggestions are provided:

- Require student machinists to read the applicable sections in the text and then complete a facts sheet about the operation.
- Guide beginning student machinists through each competency development activity by first presenting a lecture/demonstration that verbally conveys the concepts, background information, procedures, and techniques particular to that operation. This presentation should also include a live demonstration so students can observe the actual operation.
- Provide students an appropriate amount of time and material to practice the operation under instructor guidance. Provide constructive feedback. If a student is found to be lacking a complete understanding of the operation being practiced, the student should review the referenced sections in the ***Machining Fundamentals*** textbook. This review should be done before the student receives any additional information from the instructor. This requirement serves to reinforce the idea that students should, at least in some measure, be responsible for their own learning. Such a requirement enhances their ability to use the resources available to them—most notably their textbook.
- Conduct a performance test in which the student must demonstrate mastery of the operation by machining the competency piece under test conditions (no assistance from instructor or classmates). This will allow the instructor to objectively assess what the student actually knows and can do.

- Use the built-in flexibility of this book by assigning remediation work to individuals or groups if mastery of a particular operation cannot be demonstrated. This same flexibility can be used to assign enrichment activities for those who excel.
- The *Projects* section of this book contains a wide selection of advanced projects. These projects provide the opportunity for application of the skills attained in the competency development sequences. These projects afford both the instructor and the student a wide array of choices for advanced work. They may be assigned as facility and material resources permit.

## To the Student Machinist

This book is designed to guide students through the process of mastering the concepts, techniques, and procedures that all machinists must know and be able to do in order to gain employment and advance in the machining industry. Everything in this book has one ultimate purpose: to help students become competent machinists by the time they complete their education and move into the working world. It is the goal that students become immediately productive upon employment in a machining position and accustomed to seeking out information and solving problems. Those two attributes form the core of the body of skills and knowledge that will help students advance to higher levels of craftsmanship and responsibility. Listed below are some suggestions for the effective use of this book.

- **Read the textbook.** The *Machining Fundamentals* textbook is the single most important source of information about machining that you have. It was written by experts in the machining industry and has been reviewed by many industry professionals. Learn to locate needed information in the textbook. Then, apply that information to help perform a task or find a solution to a machining problem.
- **Use the orders of operations.** Presented with the competency piece or project drawing, students will find an order of operations that is intended to serve as a guide in the mastery of a particular competency or completion of a project. In the competency development sections, the orders of operations have references to the sections in the *Machining Fundamentals* textbook that cover those particular operations. The operations have been sequenced in such a way as to allow the completion of the piece in the most efficient way possible.
- **Take notes.** After each order of operation section, students will find a space reserved specifically for notes. This space is provided so students may write down important details about the operations that must be performed to complete the piece. By taking notes on the material read, students are creating a handy reference source that can be easily reviewed while working in the lab.

## Shop Safety

In any activity that involves the use of tools and machinery, the subject of safety is critically important. This is true from the standpoint of the individual machinist but also as it relates to insurance costs for the company and ultimately to the cost of the products being manufactured. The production time lost to accidents and injuries could significantly impact a company's ability to stay in business. A serious injury can also have a devastating effect on the injured person and their family. With these things in mind, study and follow all safety rules related to the tools and machines required to complete the projects contained in this book.

### General Shop Safety

✓ Keep the shop clean. Metal scraps should be placed in the scrap bin.
✓ Exercise extreme care when machining unfamiliar materials. Do not machine a material until the material is identified and procedures for handling it are known.
✓ If ill and taking medication, check with a doctor or the school clinic to determine whether it is safe to operate machinery.

- ✓ Avoid using compressed air to remove chips and cutting oil from machines.
- ✓ Oily rags must be placed in an approved safety container (a metal can with a lid).
- ✓ Use care when handling long pieces of metal stock. The end could come into contact with electrical outlets or fixtures and cause electrocution. Additionally, other workers could be struck and injured.
- ✓ Dress properly for working around machinery. Baggy clothes and long hair can become entangled in moving machine parts and cause injury or death.
- ✓ Wear appropriate safety equipment (safety glasses or goggles, gloves, steel-toed shoes, face shields, helmets, etc.)
- ✓ If what must be done or how a task should be performed is not clear, get help.

## General Machine Safety

- ✓ Never operate a machine until all guards are in place.
- ✓ Always stop a machine to make adjustments or measurements.
- ✓ Never attempt to touch the surface of a workpiece while it is in motion.
- ✓ Keep the floor around a machine clear of oil, chips, and metal scrap.
- ✓ It is considered unsafe practice to talk to others while operating a machine.
- ✓ Never attempt to remove chips or cuttings with hands or while the machine is in operation.
- ✓ Never carry sharp-pointed tools in pockets.
- ✓ Make sure tools are sharpened, in good condition, and fitted with suitable handles.
- ✓ Secure prompt medical attention for any cut, bruise, scratch, burn, or other injury. No matter how minor the injury may appear, report it to the instructor, as the threat of infection is always present.

## Bench Work Safety

- ✓ Never use a screwdriver as a substitute for a chisel.
- ✓ Wear safety goggles when regrinding screwdriver tips.
- ✓ Avoid carrying screwdrivers in pockets.
- ✓ Never strike two hammers together. The blow may cause pieces of the hardened faces to fly off and cause injury.
- ✓ Do not use a hammer unless the head is on tightly and the handle is in good condition.
- ✓ Do not "choke up" too far on the handle when striking a blow.
- ✓ When using a chisel, wear safety goggles and erect a shield around the work to protect others.
- ✓ Hold a chisel in a manner so as to prevent hand injury if it is missed with the hammer.
- ✓ If the head of a chisel becomes "mushroomed," remove the excess metal by grinding.
- ✓ Edges of metal cut with a chisel are sharp and can cause bad cuts. Remove them by grinding or filing.
- ✓ Never test the sharpness of a hacksaw blade by running fingers across its teeth.
- ✓ Store hacksaws in a way that prevents accidentally grasping the teeth when picking it up.
- ✓ Burrs formed on the cut edge of metal are sharp and can cause a serious cut.
- ✓ Do not use the hand to brush away chips; always use a chip brush.
- ✓ Always wear safety glasses or goggles while using a hacksaw. All-hard blades can shatter and produce flying chips.
- ✓ Be sure the hacksaw blade is properly tensioned.

- ✓ Never use a file that does not have a handle.
- ✓ Clean files with a file card.
- ✓ Do not try to clean a file by slapping it on a bench, as it may shatter.
- ✓ Files are very brittle. Never use one for prying tasks.
- ✓ Use a piece of cloth, not bare hands, to clean the surface being filed.
- ✓ Never hammer on or with a file.
- ✓ Chips produced by hand threading are sharp. Use a brush or piece of cloth to remove them.
- ✓ Newly cut external threads are very sharp.
- ✓ Broken taps have very sharp edges and are dangerous. Handle them with care.
- ✓ Avoid rubbing fingers or hands over polished surfaces or surfaces to be polished.
- ✓ Wash hands thoroughly after polishing operations.

## Grinding Safety

- ✓ Wear safety glasses and be sure the grinder eye shield is in place before doing any grinding.
- ✓ Do not make tool rest adjustments while the grinding wheel or belt is revolving.
- ✓ Never attempt to operate any grinding machine while taking medication or other substances that may impair your senses.
- ✓ Be sure all wheel guards and safety devices are in place before attempting to use a grinder or abrasive belt machine.
- ✓ Stand to one side of the machine during operation. Do not stand directly in front of the wheel.
- ✓ Hold small work in a clamp or hand vise. Under no condition should work be held with a cloth.
- ✓ Do not apply side pressure to a grinding wheel.
- ✓ Keep your hands clear of the rotating wheel.
- ✓ Never operate a grinding wheel at speeds greater than those recommended by the manufacturer.
- ✓ Allow the wheels or belt to stop completely before attempting to make any machine adjustments.

## Sawing and Cutoff Machine Safety

- ✓ Never attempt to operate a sawing machine while taking medication or other substances that may impair your senses.
- ✓ Get help when lifting and cutting heavy material.
- ✓ Clean oil, grease, and coolant from the floor around the work area.
- ✓ Burrs on cut pieces are sharp. Use care when handling them.
- ✓ Handle band saw blades with extreme care. They are long and springy and can uncoil suddenly.
- ✓ Be sure all work is mounted solidly before starting a cut.
- ✓ Be sure all guards are in place before using the saw.
- ✓ Always wear a dust mask and full-face shield when cutting stock with a dry-type abrasive cutoff saw.
- ✓ Avoid standing directly in line with the blade when operating a circular cutoff saw.
- ✓ Use a brush to clean chips from the machine.
- ✓ Keep hands away from moving parts.
- ✓ Stop the machine before making adjustments.
- ✓ Have all cuts, bruises, and scratches, even minor ones, treated immediately.

*Machining Projects*

## Drill Press Safety

- ✓ Wear safety glasses or goggles when working on a drill press.
- ✓ Remove any jewelry and tuck in loose clothing so they do not become entangled in the rotating drill.
- ✓ Be sure all guards and covers are in place before operating the drill press.
- ✓ Clamp the work solidly. Serious injury can result from work that becomes loose and spins on a drill press. Never attempt to hold the work with your hand.
- ✓ When removing a drill, place a piece of wood below it. Small drills can be damaged if dropped and larger drills could cause injury.
- ✓ Never attempt to operate a drilling machine while taking medication or other substances that may impair your senses.
- ✓ Always remove the key from the chuck before turning on the power
- ✓ Let the spindle come to a stop after completing the operation. Never try to stop it with your hand.
- ✓ Clean chips from the work with a brush.
- ✓ Never attempt to use a taper shank drill bit in a drill chuck.
- ✓ Never insert a tap into the drill press and attempt to use the drill press power to run the tap into the work.

## Lathe Safety

- ✓ Never attempt to operate the lathe or any other equipment while taking medication or other substances that may impair your senses.
- ✓ Always wear safety glasses while operating a lathe.
- ✓ Dress appropriately! Remove jewelry, roll up sleeves, secure loose clothing, and tie back long hair.
- ✓ Clamp all work securely.
- ✓ Be sure all guards are in place before attempting to operate a lathe. Never attempt to defeat or bypass a safety switch.
- ✓ Turn the faceplate or chuck by hand to be sure there is no binding or danger of the work striking any part of the lathe.
- ✓ Keep the machine clear of tools and always stop the machine before making measurements or adjustments.
- ✓ Remember that metal chips are sharp and often hot. Do not try to remove them with hands. Always use pliers to clear chips from around the tool bit and holder.
- ✓ Do not permit small diameter work to project too far from the face of the chuck jaws because it could roll up over the tool bit and break.
- ✓ Do not allow the cutting tool to run into the chuck or lathe dog. This will damage the lathe and could cause injury to the operator.
- ✓ When knurling, keep the coolant brush clear of the work.
- ✓ Before repositioning or removing work from the lathe, move the cutting tool clear of the work area. This will prevent accidental cuts.
- ✓ Avoid talking while operating a lathe. Do not permit anyone else to fool around with the lathe. The operator should be the only one who turns the machine on and off or makes any adjustments.
- ✓ Never attempt to run the chuck on or off under power.
- ✓ Always remove the chuck key for the chuck before turning the machine on. Make it a practice to never let go of the chuck key until it is clear of the work area.
- ✓ When filing on the lathe, make sure the file has a securely fitted handle.
- ✓ Remove sharp edges and burrs from the workpiece before handling it.

**Milling Machine Safety**

- ✓ Always wear safety glasses and appropriate clothing when operating a milling machine.
- ✓ Never attempt to operate a milling machine or any other equipment while taking medication or other substances that may impair your senses.
- ✓ Never activate the rapid traverse while the cutter is positioned for a cut.
- ✓ Make sure the spindle has come to a complete stop before attempting to adjust V-belts.
- ✓ Because the chips produced during milling operations are sharp and often hot, always use a brush to clear them.
- ✓ Never use compressed air to blow chips off a machine.
- ✓ Always stop the machine before attempting to make adjustments or take measurements.
- ✓ Never reach over or near a rotating cutter.
- ✓ Avoid talking while operating a machine tool. Never allow anyone else to turn the machine on or off.
- ✓ Use a piece of heavy cloth or gloves for protection when handling milling cutters. Avoid using bare hands.
- ✓ Get help to move any heavy machine attachments, such as a vise, dividing head, rotary table, or large work.
- ✓ The work or work-holding device (mill vise, strap clamps, and step blocks, etc.) is locked down securely before beginning any milling operation.

Machining Projects

# Reference Material

## Decimal Equivalents: Letter-Size Drills

All decimal equivalents are rounded to the nearest thousandth according to standard mathematical practice.

| Drill | Size of Drill in Inches | Drill | Size of Drill in Inches |
|---|---|---|---|
| A | .234 | N | .302 |
| B | .238 | O | .316 |
| C | .242 | P | .323 |
| D | .246 | Q | .332 |
| E | .250 | R | .339 |
| F | .257 | S | .348 |
| G | .261 | T | .358 |
| H | .266 | U | .368 |
| I | .272 | V | .377 |
| J | .277 | W | .386 |
| K | .281 | X | .397 |
| L | .290 | Y | .404 |
| M | .295 | Z | .413 |

## Decimal Equivalents: Number-Size Drills

All decimal equivalents are rounded to the nearest thousandth according to standard mathematical practice.

| Drill | Size of Drill in Inches | Drill | Size of Drill in Inches | Drill | Size of Drill in Inches |
|---|---|---|---|---|---|
| 1 | .228 | 28 | .141 | 55 | .052 |
| 2 | .221 | 29 | .136 | 56 | .047 |
| 3 | .213 | 30 | .129 | 57 | .043 |
| 4 | .209 | 31 | .120 | 58 | .042 |
| 5 | .206 | 32 | .116 | 59 | .041 |
| 6 | .204 | 33 | .113 | 60 | .040 |
| 7 | .201 | 34 | .111 | 61 | .039 |
| 8 | .199 | 35 | .110 | 62 | .038 |
| 9 | .196 | 36 | .107 | 63 | .037 |
| 10 | .194 | 37 | .104 | 64 | .036 |
| 11 | .191 | 38 | .102 | 65 | .035 |
| 12 | .189 | 39 | .100 | 66 | .033 |
| 13 | .185 | 40 | .098 | 67 | .032 |
| 14 | .182 | 41 | .096 | 68 | .031 |
| 15 | .180 | 42 | .094 | 69 | .029 |
| 16 | .177 | 43 | .089 | 70 | .028 |
| 17 | .173 | 44 | .086 | 71 | .026 |
| 18 | .170 | 45 | .082 | 72 | .025 |
| 19 | .166 | 46 | .081 | 73 | .024 |
| 20 | .161 | 47 | .079 | 74 | .023 |
| 21 | .159 | 48 | .076 | 75 | .021 |
| 22 | .157 | 49 | .073 | 76 | .020 |
| 23 | .154 | 50 | .070 | 77 | .018 |
| 24 | .152 | 51 | .067 | 78 | .016 |
| 25 | .150 | 52 | .064 | 79 | .015 |
| 26 | .147 | 53 | .060 | 80 | .014 |
| 27 | .144 | 54 | .055 | | |

Copyright by Goodheart-Willcox Co., Inc.

The following table is a starting point for drilling different materials. Feeds and speeds should be increased and decreased depending upon the specific metal being drilled and the condition of the drill press. Additional information on drill speeds can be found in the *Machinery's Handbook*.

| Diameter of drill | Soft metals 300 fpm | Plastics and hard rubber 200 fpm | Annealed cast from 140 fpm | Mild steel 100 fpm | Malleable iron 90 fpm | Hard cast iron 80 fpm | Tool or hard steel 60 fpm | Alloy steel cast steel 40 fpm |
|---|---|---|---|---|---|---|---|---|
| $1/16$ (No. 53 to 80) | 18320 | 12217 | 8554 | 6111 | 5500 | 4889 | 3667 | 2445 |
| $3/32$ (No. 42 to 52) | 12212 | 8142 | 5702 | 4071 | 3666 | 3258 | 2442 | 1649 |
| $1/8$ (No. 31 to 41) | 9160 | 6112 | 4278 | 3056 | 2750 | 2445 | 1833 | 1222 |
| $5/32$ (No. 23 to 30) | 7328 | 4888 | 3420 | 2444 | 2198 | 1954 | 1465 | 977 |
| $3/16$ (No. 13 to 22) | 6106 | 4075 | 2852 | 2037 | 1833 | 1630 | 1222 | 815 |
| $7/32$ (No. 1 to 12) | 5234 | 3490 | 2444 | 1745 | 1575 | 1396 | 1047 | 698 |
| $1/4$ (A to E) | 4575 | 3055 | 2139 | 1527 | 1375 | 1222 | 917 | 611 |
| $9/32$ (G to K) | 4071 | 2712 | 1900 | 1356 | 1222 | 1084 | 814 | 542 |
| $5/16$ (L, M, N) | 3660 | 2445 | 1711 | 1222 | 1100 | 978 | 733 | 489 |
| $11/32$ (O to R) | 3330 | 2220 | 1554 | 1110 | 1000 | 888 | 666 | 444 |
| $3/8$ (S, T, U) | 3050 | 2037 | 1426 | 1018 | 917 | 815 | 611 | 407 |
| $13/32$ (V to Z) | 2818 | 1878 | 1316 | 939 | 846 | 752 | 563 | 376 |
| $7/16$ | 2614 | 1746 | 1222 | 873 | 786 | 698 | 524 | 349 |
| $15/32$ | 2442 | 1628 | 1140 | 814 | 732 | 652 | 488 | 326 |
| $1/2$ | 2287 | 1528 | 1070 | 764 | 688 | 611 | 458 | 306 |
| $9/16$ | 2035 | 1357 | 950 | 678 | 611 | 543 | 407 | 271 |
| $5/8$ | 1830 | 1222 | 856 | 611 | 550 | 489 | 367 | 244 |
| $11/16$ | 1665 | 1110 | 777 | 555 | 500 | 444 | 333 | 222 |
| $3/4$ | 1525 | 1018 | 713 | 509 | 458 | 407 | 306 | 204 |

Figures are for high-speed drills. The speed of carbon drills should be reduced one-half. Use drill speed nearest to figure given.

## Machining Projects

### Cutting Speeds for Round Stock

| Diameter | Material | Roughing Cut (rpm) | Finishing Cut (rpm) | Threading (rpm) | Diameter | Material | Roughing Cut (rpm) | Finishing Cut (rpm) | Threading (rpm) |
|---|---|---|---|---|---|---|---|---|---|
| 1/8" | Machine steel/bronze<br>Cast iron<br>Tool steel (annealed)<br>Brass<br>Aluminum | 2880<br>1920<br>1600<br>4800<br>6400 | 3200<br>2560<br>2400<br>6400<br>9600 | 1020<br>800<br>640<br>1600<br>1600 | 1 1/2" | Machine steel/bronze<br>Tool steel<br>Brass<br>Aluminum | 240<br>134<br>400<br>534 | 270<br>200<br>534<br>800 | 67<br>53<br>134<br>134 |
| 3/16" | Machine steel/bronze<br>Tool steel<br>Brass<br>Aluminum | 2880<br>1600<br>4800<br>6400 | 3200<br>2400<br>6400<br>9600 | 1120<br>640<br>1600<br>1600 | 1 3/4" | Machine steel/bronze<br>Tool steel<br>Brass<br>Aluminum | 205<br>115<br>340<br>456 | 230<br>170<br>450<br>680 | 80<br>50<br>115<br>115 |
| 1/4" | Machine steel/bronze<br>Tool steel<br>Brass<br>Aluminum | 1440<br>800<br>2400<br>3200 | 1600<br>1200<br>3200<br>4800 | 560<br>320<br>800<br>800 | 2" | Machine steel<br>Tool steel<br>Brass<br>Aluminum | 180<br>100<br>300<br>400 | 200<br>150<br>400<br>600 | 50<br>40<br>100<br>100 |
| 3/8" | Machine steel/bronze<br>Tool steel<br>Brass<br>Aluminum | 960<br>540<br>1700<br>2130 | 1066<br>800<br>2100<br>3200 | 270<br>220<br>530<br>540 | 2 1/2" | Machine steel<br>Tool steel<br>Brass<br>Aluminum | 141<br>80<br>240<br>320 | 160<br>120<br>320<br>480 | 56<br>32<br>80<br>80 |
| 1/2" | Machine steel/bronze<br>Tool steel<br>Brass<br>Aluminum | 720<br>400<br>1200<br>1600 | 800<br>600<br>1600<br>2400 | 280<br>160<br>400<br>400 | 3" | Machine steel<br>Tool steel<br>Brass<br>Aluminum | 120<br>65<br>200<br>270 | 140<br>100<br>270<br>400 | 40<br>40<br>65<br>65 |
| 5/8" | Machine steel/bronze<br>Tool steel<br>Brass<br>Aluminum | 576<br>320<br>960<br>1280 | 640<br>480<br>1280<br>1920 | 160<br>200<br>320<br>320 | 3 1/2" | Machine steel<br>Tool steel<br>Brass<br>Aluminum | 103<br>60<br>171<br>228 | 115<br>85<br>228<br>342 | 40<br>23<br>57<br>57 |
| 3/4" | Machine steel/bronze<br>Tool steel<br>Brass<br>Aluminum | 500<br>266<br>800<br>1066 | 550<br>400<br>1066<br>1600 | 176<br>106<br>266<br>266 | 4" | Machine steel<br>Tool steel<br>Brass<br>Aluminum | 90<br>50<br>150<br>200 | 100<br>75<br>200<br>300 | 35<br>20<br>50<br>50 |
| 1" | Machine steel/bronze<br>Tool steel<br>Brass<br>Aluminum | 360<br>200<br>600<br>800 | 400<br>300<br>800<br>1200 | 140<br>80<br>200<br>200 | 4 1/2" | Machine steel<br>Tool steel<br>Brass<br>Aluminum | 80<br>45<br>133<br>178 | 90<br>67<br>178<br>267 | 31<br>18<br>45<br>45 |
| 1 1/4" | Machine steel<br>Tool steel<br>Brass<br>Aluminum | 288<br>160<br>480<br>640 | 320<br>240<br>640<br>960 | 112<br>64<br>160<br>160 | 5" | Machine steel<br>Tool steel<br>Brass<br>Aluminum | 72<br>40<br>120<br>160 | 80<br>58<br>160<br>240 | 28<br>16<br>40<br>40 |

## Rules for Determining Speeds and Feeds

| To Find | Having | Rule | Formula |
|---|---|---|---|
| Speed of cutter in feet per minute (FPM) | Diameter of cutter and revolutions per minute | Diameter of cutter (in inches) multiplied by 3.1416 ($\pi$) multiplied by revolutions per minute, divided by 12 | $FPM = \dfrac{\pi D \times RPM}{12}$ |
| Speed of cutter in meters per minute (MPM) | Diameter of cutter and revolutions per minute | Diameter of cutter multiplied by 3.1416 ($\pi$) multiplied by revolutions per minute, divided by 1000 | $MPM = \dfrac{D(mm) \times \pi \times RPM}{1000}$ |
| Revolutions per minute (RPM) | Feet per minute and diameter of cutter | Feet per minute, multiplied by 12, divided by circumference of cutter ($\pi D$) | $RPM = \dfrac{FPM \times 12}{\pi D}$ |
| Revolutions per minute (RPM) | Meters per minute and diameter of cutter in millimeters (mm) | Meters per minute, multiplied by 1000, divided by the circumference of cutter ($\pi D$) | $RPM = \dfrac{MPM \times 1000}{\pi D}$ |
| Feed per revolution (FR) | Feed per minute and revolutions per minute | Feed per minute, divided by revolutions per minute | $FR = \dfrac{F}{RPM}$ |
| Feed per tooth per revolution (FTR) | Feed per minute and number of teeth in cutter | Feed per minute (in inches or millimeters) divided by number of teeth in cutter $\times$ revolutions per minute | $FTR = \dfrac{F}{T \times RPM}$ |
| Feed per minute (F) | Feed per tooth per revolution, number of teeth in cutter, and RPM | Feed per tooth per revolutions multiplied by number of teeth in cutter, multiplied by revolutions per minute | $F = FTR \times T \times RPM$ |
| Feed per minute (F) | Feed per revolution and revolutions per minute | Feed per revolution multiplied by revolutions per minute | $F = FR \times RPM$ |
| Number of teeth per minute (TM) | Number of teeth in cutter and revolutions per minute | Number of teeth in cutter multiplied by revolutions per minute | $TM = T \times RPM$ |

RPM = Revolutions per minute  
T = Teeth in cutter  
D = Diameter of cutter  
$\pi$ = 3.1416 (pi)  
FPM = Speed of cutter in feet per minute  
TM = Teeth per minute  
F = Feed per minute  
FR = Feed per revolution  
FTR = Feed per tooth per revolution  
MPM = Speed of cutter in meters per minute

## Recommended Turning Rates for Stainless Steels Using High-speed Tools

| Nature of Stock | Type No. | Speed (sfpm) | Feed (inches per revolution) |
|---|---|---|---|
| Free machining grades | 430 F | 100 – 140 | 0.003-0.005 for finish cuts and up to 0.015 for roughing cuts |
|  | 416 | 90 – 135 |  |
|  | 303 | 80 – 120 |  |
| High-carbon grades that are slowed down due to their abrasive action on tools | 410 | 75 – 115 | 0.003–0.008 |
|  | 430 | 75 – 115 | 0.003–0.008 |
|  | 420 | 45 – 85 | 0.003–0.008 |
|  | 431 | 45 – 85 | 0.003–0.008 |
|  | 440 | 30 – 60 | 0.003–0.008 |
|  | 302 |  |  |
|  | 304 |  |  |
|  | 316 | 45 – 80 | 0.004–0.008 |

## Machining Projects

### Feeds and Speeds for HSS Drills, Reamers, and Taps

| Material | Brinell | Drills (sfm) | Point | Feed | Reamers (sfm) | Feed | Taps (sfm) Threads per Inch 3–7 1/2 | 8–15 | 16–24 | 25–up |
|---|---|---|---|---|---|---|---|---|---|---|
| Aluminum | 99–101 | 200–250 | 118° | M | 150–160 | M | 50 | 100 | 150 | 200 |
| Aluminum bronze | 170–187 | 60 | 118° | M | 40–45 | M | 12 | 25 | 45 | 60 |
| Bakelite | ... | 80 | 60°-90° | M | 50–60 | M | 50 | 100 | 150 | 200 |
| Brass | 192–202 | 200–250 | 118° | H | 150–160 | H | 50 | 100 | 150 | 200 |
| Bronze, common | 166–183 | 200–250 | 118° | H | 150–160 | H | 40 | 80 | 100 | 150 |
| Bronze, phosphor, 1/2 hard | 187–202 | 175–180 | 118° | M | 130–140 | M | 25 | 40 | 50 | 80 |
| Bronze, phosphor, soft | 149–163 | 200–250 | 118° | H | 150–160 | H | 40 | 80 | 100 | 150 |
| Cast iron, soft | 126 | 140–150 | 90° | H | 100–110 | H | 30 | 60 | 90 | 140 |
| Cast iron, medium soft | 196 | 80–110 | 118° | M | 50–65 | M | 25 | 40 | 50 | 80 |
| Cast iron, hard | 293–302 | 45–50 | 118° | L | 67–75 | L | 10 | 20 | 30 | 40 |
| Cast iron, chilled* | 402 | 15 | 150° | L | 8–10 | L | 5 | 5 | 10 | 10 |
| Cast steel | 286–302 | 40–50* | 118° | L | 70–75 | L | 20 | 30 | 40 | 50 |
| Celluloid | ... | 100 | 90° | M | 75–80 | M | 50 | 100 | 150 | 200 |
| Copper | 80–85 | 70 | 100° | L | 45–55 | L | 40 | 80 | 100 | 150 |
| Drop forgings (steel) | 170–196 | 60 | 118° | M | 40–45 | M | 12 | 25 | 45 | 60 |
| Duralumin | 90–104 | 200 | 118° | M | 150–160 | M | 50 | 100 | 150 | 200 |
| Everdur | 179–207 | 60 | 118° | L | 40–45 | L | 20 | 30 | 40 | 50 |
| Machinery steel | 170–196 | 110 | 118° | H | 67–75 | H | 35 | 50 | 60 | 85 |
| Magnet steel, soft | 241–302 | 35–40 | 118° | M | 20–25 | M | 20 | 40 | 50 | 75 |
| Magnet steel, hard* | 321–512 | 15 | 150° | L | 10 | L | 5 | 10 | 15 | 25 |
| Manganese steel, 7% – 13% | 187–217 | 15 | 150° | L | 10 | L | 15 | 20 | 25 | 30 |
| Manganese copper, 30% Mn.* | 134 | 15 | 150° | L | 10–12 | L | ... | ... | ... | ... |
| Malleable iron | 112–126 | 85–90 | 118° | H | ... | H | 20 | 30 | 40 | 50 |
| Mild steel, .20 –.30 C | 170–202 | 110–120 | 118° | H | 75–85 | H | 40 | 55 | 70 | 90 |
| Molybdenum steel | 196–235 | 55 | 125° | M | 35–45 | M | 20 | 30 | 35 | 45 |
| Monel metal | 149–170 | 50 | 118° | M | 35–38 | M | 8 | 10 | 15 | 20 |
| Nickel, pure* | 187–202 | 75 | 118° | L | 40 | L | 25 | 40 | 50 | 80 |
| Nickel steel, 3 1/2% | 196–241 | 60 | 118° | L | 40–45 | L | 8 | 10 | 15 | 20 |
| Rubber, hard | ... | 100 | 60°-90° | L | 70–80 | L | 50 | 100 | 150 | 200 |
| Screw stock, C.R. | 170–196 | 110 | 118° | H | 75 | H | 20 | 30 | 40 | 50 |
| Spring steel | 402 | 20 | 150° | L | 12–15 | L | 10 | 10 | 15 | 15 |
| Stainless steel | 146–149 | 50 | 118° | M | 30 | M | 8 | 10 | 15 | 20 |
| Stainless steel, C.R.* | 460–477 | 20 | 118° | L | 15 | L | 8 | 10 | 15 | 20 |
| Steel, .40 to .50 C | 170–196 | 80 | 118° | M | 8–10 | M | 20 | 30 | 40 | 50 |
| Tool, S.A.E., and forging steel | 149 | 75 | 118° | H | 35–40 | H | 25 | 35 | 45 | 55 |
| Tool, S.A.E., and forging steel | 241 | 50 | 125° | M | 12 | M | 15 | 15 | 25 | 25 |
| Tool, S.A.E., and forging steel* | 402 | 15 | 150° | L | 10 | L | 8 | 10 | 15 | 20 |
| Zinc alloy | 112–126 | 200–250 | 118° | M | 150–175 | M | 50 | 100 | 150 | 200 |

*Use specially constructed heavy-duty drills.
Note: Carbon steel tools should be run at speeds 40% to 50% of those recommended for high speed steel.
Spiral point taps may be run at speeds 15% to 20% faster than regular taps.

## Feeds and Speeds for HSS Drills in Various Metals

| Drill diameter | Cast iron | | Bronze or Brass | | Drop forgings Alloy steel Tool steel Annealed | | Drop forgings Alloy steel Heat-treated | | Steel castings | | Mild steel | |
|---|---|---|---|---|---|---|---|---|---|---|---|---|
| | Feed | Speed | Feed | Speed | Feed | Speed | Feed | Speed | Feed | Speed | Feed | Speed |
| 1/16" | .002 | 4550 | .002 | 9150 | .002 | 3650 | .002 | 2750 | .002 | 3650 | .002 | 4250 |
| | .004 | 6700 | .004 | 12,000 | .003 | 4550 | .003 | 3650 | .003 | 4550 | .003 | 5600 |
| 1/8" | .002 | 2550 | .002 | 4550 | .002 | 1800 | .002 | 1225 | .002 | 1800 | .002 | 2100 |
| | .004 | 3350 | .004 | 5600 | .003 | 2250 | .003 | 1800 | .003 | 2250 | .003 | 2800 |
| 3/16" | .004 | 1500 | .004 | 3100 | .003 | 1200 | .003 | 900 | .003 | 1200 | .003 | 1400 |
| | .006 | 2200 | .007 | 5600 | .004 | 1500 | .004 | 1200 | .005 | 1500 | .005 | 1900 |
| 1/4" | .004 | 1150 | .004 | 2300 | .003 | 925 | .003 | 750 | .003 | 925 | .003 | 1050 |
| | .006 | 1650 | .007 | 2750 | .004 | 1150 | .004 | 925 | .005 | 1150 | .005 | 1500 |
| 5/16" | .006 | 925 | .007 | 1825 | .004 | 725 | .004 | 500 | .004 | 725 | .005 | 850 |
| | .009 | 1325 | .010 | 2200 | .006 | 925 | .005 | 725 | .006 | 925 | .007 | 1200 |
| 3/8" | .006 | 750 | .007 | 1525 | .004 | 600 | .004 | 400 | .004 | 600 | .005 | 700 |
| | .009 | 1100 | .010 | 1850 | .006 | 750 | .005 | 600 | .006 | 750 | .007 | 925 |
| 7/16" | .009 | 650 | .010 | 1300 | .006 | 525 | .005 | 350 | .006 | 525 | .006 | 600 |
| | .012 | 950 | .014 | 1525 | .009 | 650 | .006 | 525 | .010 | 650 | .010 | 800 |
| 1/2" | .008 | 575 | .010 | 1150 | .006 | 375 | .005 | 300 | .006 | 375 | .006 | 525 |
| | .012 | 850 | .014 | 1375 | .009 | 575 | .006 | 375 | .010 | 575 | .010 | 700 |
| 9/16" | .012 | 500 | .014 | 1000 | .008 | 350 | .007 | 275 | .010 | 350 | .010 | 575 |
| | .016 | 750 | .018 | 1200 | .012 | 500 | .010 | 350 | .014 | 500 | .014 | 625 |
| 5/8" | .012 | 450 | .014 | 900 | .008 | 300 | .007 | 250 | .010 | 300 | .010 | 425 |
| | .016 | 675 | .018 | 1100 | .012 | 450 | .010 | 300 | .014 | 450 | .014 | 565 |
| 11/16" | .012 | 410 | .014 | 800 | .008 | 275 | .007 | 225 | .010 | 275 | .010 | 375 |
| | .016 | 625 | .018 | 1000 | .012 | 410 | .010 | 275 | .014 | 410 | .014 | 525 |
| 3/4" | .012 | 375 | .014 | 750 | .008 | 250 | .007 | 200 | .010 | 250 | .010 | 350 |
| | .016 | 550 | .018 | 900 | .012 | 375 | .010 | 250 | .014 | 375 | .014 | 475 |
| 13/16" | .014 | 350 | .016 | 700 | .010 | 240 | .009 | 190 | .014 | 240 | .014 | 325 |
| | .020 | 525 | .022 | 850 | .014 | 350 | .012 | 240 | .016 | 350 | .016 | 450 |
| 7/8" | .014 | 325 | .016 | 650 | .010 | 225 | .009 | 175 | .014 | 225 | .014 | 300 |
| | .020 | 475 | .022 | 800 | .014 | 325 | .012 | 225 | .016 | 325 | .016 | 400 |
| 15/16" | .014 | 300 | .016 | 625 | .010 | 200 | .009 | 160 | .014 | 200 | .014 | 275 |
| | .020 | 450 | .022 | 725 | .014 | 300 | .012 | 200 | .016 | 300 | .016 | 375 |
| 1" | .014 | 280 | .016 | 575 | .010 | 185 | .009 | 150 | .014 | 185 | .014 | 265 |
| | .020 | 425 | .022 | 675 | .014 | 280 | .012 | 185 | .016 | 280 | .016 | 350 |

(Chicago-Latrobe)

Speeds and feeds shown apply to average working conditions and materials. They are recommended with regard to conserving drills and avoiding excessive machine tool wear. Under many conditions, these speeds and feeds may be considerably increased; under others they must be decreased. This is dependent on judgment of operator and performance obtained. Excessive speeds and feeds will show up by action of machine and drill. Same applies to lower speeds and feeds. Operator will notice whether he/she is getting proper performance by experience, and will advance or retard as case may justify. Feeds and speeds should be changed in proper proportions and a liberal use of cooling compound will increase life of tools.

Never dip a drill into water to cool it while grinding. This will cause tiny checks, or cracks at the cutting edge, which will cause the drill to dull quickly.

Do not leave a drill in after it shows signs of dulling or laboring; then is the time to regrind. Proper grinding is essential.

*To determine feed and speed according to the above chart, proceed as follows:*

You are going to drill heat-treated drop forgings. We suppose you will use a 1/2" drill. Follow column down to where the 1/2" drill meets it; there you will find that a feed from .005 to .006 and a speed of from 300 to 375 rpm are recommended. Start by using .005 feed and 300 rpm. If drill and machine seem to turn smoothly without strain, then both feed and speed can be advanced. Operator will soon find which is best.

## Machining Projects

| Formulas for Machining Bar Stock | |
|---|---|
| Surface speed—feet/minute | |
|     Round bars | $\dfrac{\text{Diameter} \times 3.1416 \times \text{rpm}}{12}$ |
|     Hexagon bars (distance across corners) | $\dfrac{\text{Size} \times 3.1416 \times \text{rpm} \times 1.155}{12}$ |
| | $\dfrac{\text{Distance} \times 3.1416 \times \text{rpm}}{12}$ |
|     Square bars (distance across corners) | $\dfrac{\text{Size} \times 3.1416 \times \text{rpm} \times 1.414}{12}$ |
| | $\dfrac{\text{Distance} \times 3.1416 \times \text{rpm}}{12}$ |
| Revolutions—number/minute | |
|     Round bars | $\dfrac{\text{SFM} \times 12}{\text{Diameter} \times 3.1416}$ |
|     Hexagon bars | $\dfrac{\text{SFM} \times 12}{\text{Size} \times 3.1416 \times 1.155}$ |
| | $\dfrac{\text{SFM} \times 12}{\text{Distance} \times 3.1416}$ |
|     Square bars | $\dfrac{\text{SFM} \times 12}{\text{Size} \times 3.1416 \times 1.414}$ |
| | $\dfrac{\text{SFM} \times 12}{\text{Distance} \times 3.1416}$ |
| Feed—inches/revolution | $\dfrac{\text{Feed inches per minute}}{\text{rpm}}$ |
| | $\dfrac{\text{Diameter} \times 3.1416 \times \text{Feed}}{\text{SFM} \times 12}$ |
| Feed—inches/tooth | $\dfrac{\text{Feed}}{\text{Number of teeth}}$ |
| Time for actual machining—seconds | $\dfrac{\text{Revolutions required} \times 60 \text{ seconds}}{\text{rpm}}$ |
| Machine time | Time for machining + idle time |
| Tapping or threading time—seconds | $\dfrac{\text{Number of threads} \times 60 \text{ seconds}}{\text{Actual threading speed in rpm}}$ |

| Cutting Fluids for Various Metals | |
|---|---|
| Aluminum and Its Alloys | Kerosene, kerosene and lard oil, soluble oil |
| Plastics | Dry |
| Brass, Soft | Dry, soluble oil, kerosene and lard oil |
| Bronze, High Tensile | Soluble oil, lard oil, mineral oil, dry |
| Cast Iron | Dry, air jet, soluble oil |
| Copper | Soluble oil, dry, mineral lard oil, kerosene |
| Magnesium | Low viscosity neutral oils |
| Malleable Iron | Dry, soda water |
| Monel Metal | Lard oil, soluble oil |
| Slate | Dry |
| Steel, Forging | Soluble oil, sulfurized oil, mineral lard oil |
| Steel, Manganese | Soluble oil, sulfurized oil, mineral lard oil |
| Steel, Soft | Soluble oil, mineral lard oil, sulfurized oil, lard oil |
| Steel, Stainless | Sulfurized mineral oil, soluble oil |
| Steel, Tool | Soluble oil, mineral lard oil, sulfurized oil |
| Wrought Iron | Soluble oil, mineral lard oil, sulfurized oil |

# Rules for Figuring Tapers

D = Diameter at large end
d = Diameter at small end
$\ell$ = Length of taper
L = Total length of piece
TPI = Taper per inch
TPF = Taper per foot

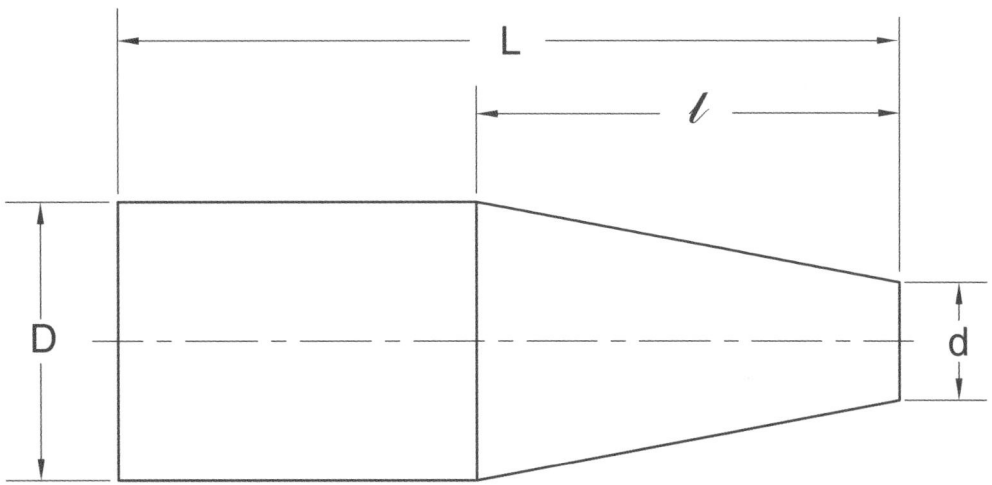

## Calculating Offset When Taper Per Inch Is Known

Information needed:
TPI = Taper per inch
L = Total length of piece
Formula used: L × TPI / 2
What will be the tailstock offset for a piece with a total length of 8″ and a taper per inch of 0.125″?
Taper per inch        = 0.125
Total length of piece = 8
Offset                = L × TPI / 2
                      = 8 × 0.125 / 2
                      = 0.5″

## Calculating Offset When Taper Per Foot Is Known

Information needed:
TPF = Taper per foot
L = Total length of piece
Formula used: L × TPF / 24
What will be the tailstock offset for a piece with a total length of 8″ and a taper per foot of 1.5″?
Taper per foot        = 1.5
Total length of piece = 8
Offset                = L × TPF / 24
                      = 8 × 1.5 / 24
                      = 0.5″

## Calculating Offset When Dimension of Taper Sections Are Known but TPI or TPF Is Not Known

Information needed:
D = Diameter at large end
d = Diameter at small end
$l$ = Length of taper
L = Total length of piece
Formula used: $\dfrac{L \times (D-d)}{2l}$

What will be the tailstock offset for a 9″ piece with a 1.250″ large end diameter, 0.875″ small end diameter, and a 3″ taper length?

Diameter at large end = 1.250″
Diameter at small end = 0.875″
Length of taper = 3″
Total length of piece = 9″

$$\text{Offset} = \dfrac{L \times (D-d)}{2l}$$

$$= \dfrac{9 \times (1.250 - 0.875)}{2(3.000)}$$

$$= \dfrac{9 \times 0.375}{6}$$

$$= 0.562″$$

| Taper per Foot with Corresponding Angles | | |
|---|---|---|
| Taper per foot | Included angle | Angle with centerline |
| 1/16 | 0° 17′ 53″ | 0° 8′ 57″ |
| 1/8 | 0° 35′ 47″ | 0° 17′ 54″ |
| 3/16 | 0° 53′ 44″ | 0° 26′ 52″ |
| 1/4 | 1° 11′ 38″ | 0° 35′ 49″ |
| 5/16 | 1° 29′ 31″ | 0° 44′ 46″ |
| 3/8 | 1° 47′ 25″ | 0° 53′ 42″ |
| 7/16 | 2° 5′ 18″ | 1° 2′ 39″ |
| 1/2 | 2° 23′ 12″ | 1° 11′ 36″ |
| 9/16 | 2° 41′ 7″ | 1° 20′ 34″ |
| 5/8 | 2° 58′ 3″ | 1° 29′ 31″ |
| 11/16 | 3° 16′ 56″ | 1° 38′ 28″ |
| 3/4 | 3° 34′ 48″ | 1° 47′ 24″ |
| 13/16 | 3° 52′ 42″ | 1° 56′ 21″ |
| 7/8 | 4° 10′ 32″ | 2° 5′ 16″ |
| 15/16 | 4° 28′ 26″ | 2° 14′ 13″ |
| 1 | 4° 46′ 19″ | 2° 23′ 10″ |

Machining Projects

# Section 1
## Bench Work

Project 1.1    Drill Gage

# Project 1.1 Drill Gage

Name _____  Date _____

Instructor _____  Period _____

## NIMS Duties

*The tasks in this project develop skills related to the following NIMS duties:*

- Duty 1.1   Job Process Planning
- Duty 2.1   Manual Operations: Benchwork
- Duty 2.2   Manual Operations: Layout
- Duty 2.8   Drill Press

## Order of Operations

**4.1, 12.1–12.6** ❏ 1. Cut appropriate stock 1/8" longer than the finished length.

**5.2, 6.8** ❏ 2. Using files, a precision square, and appropriate technique, square one end of the workpiece with one edge. Use a permanent marker to mark the good end and good edge.

**5.1** ❏ 3. Coat the front of the workpiece with layout dye. Be sure to have the good end and the good edge marked on the backside of the piece for reference.

**4.1, 5.1, 5.4** ❏ 4. With a combination square, dividers, protractor, and scriber, lay out all hole locations, angles, radii, and edges. Remember to take all measurements from the good end and the good edge.

**10.6, 10.9** ❏ 5. Use a wiggler/center finder to accurately locate the centers of the holes, and then drill the holes with the correct drill bits. Be sure to deburr each hole (especially on the back) so the workpiece is safe to handle.

**12.3, 12.4, 12.6** ❏ 6. Using a vertical band saw, cut away the excess material from the workpiece. Cut about 1/16" *outside* the layout lines so there is enough material to ensure that all saw marks are removed when file finishing.

**6.8** ❏ 7. Using both straight filing and draw filing techniques, file the workpiece to the specified shape. Be careful to keep all corners sharp and all edges square with the body of the piece. Keep in mind that the finished piece should look *exactly* like the drawing.

❏ 8. Stamp hole sizes beside each hole.

**6.11** ❏ 9. Using appropriate abrasives and polishing technique, polish the drill gage to a mirror finish on both sides and all edges.

Note   At a later date, the drill gage can be set up in the vertical mill and, using a carbide burr, the graduations along the cutting lip angle can be cut every 1/64" or 1/32" as desired.

## Notes

_____
_____
_____

**Performance Evaluation—Instructor's Use Only**

Project completed on time?   ❏ Yes   ❏ No
   If No, which steps were not completed _____
Overall performance on project:
   ❏ Excellent   ❏ Good   ❏ Satisfactory   ❏ Unsatisfactory   ❏ Poor
Comments: _____
_____

Instructor's Signature _____

*Machining Projects*  **Project 1.1    Drill Gage**

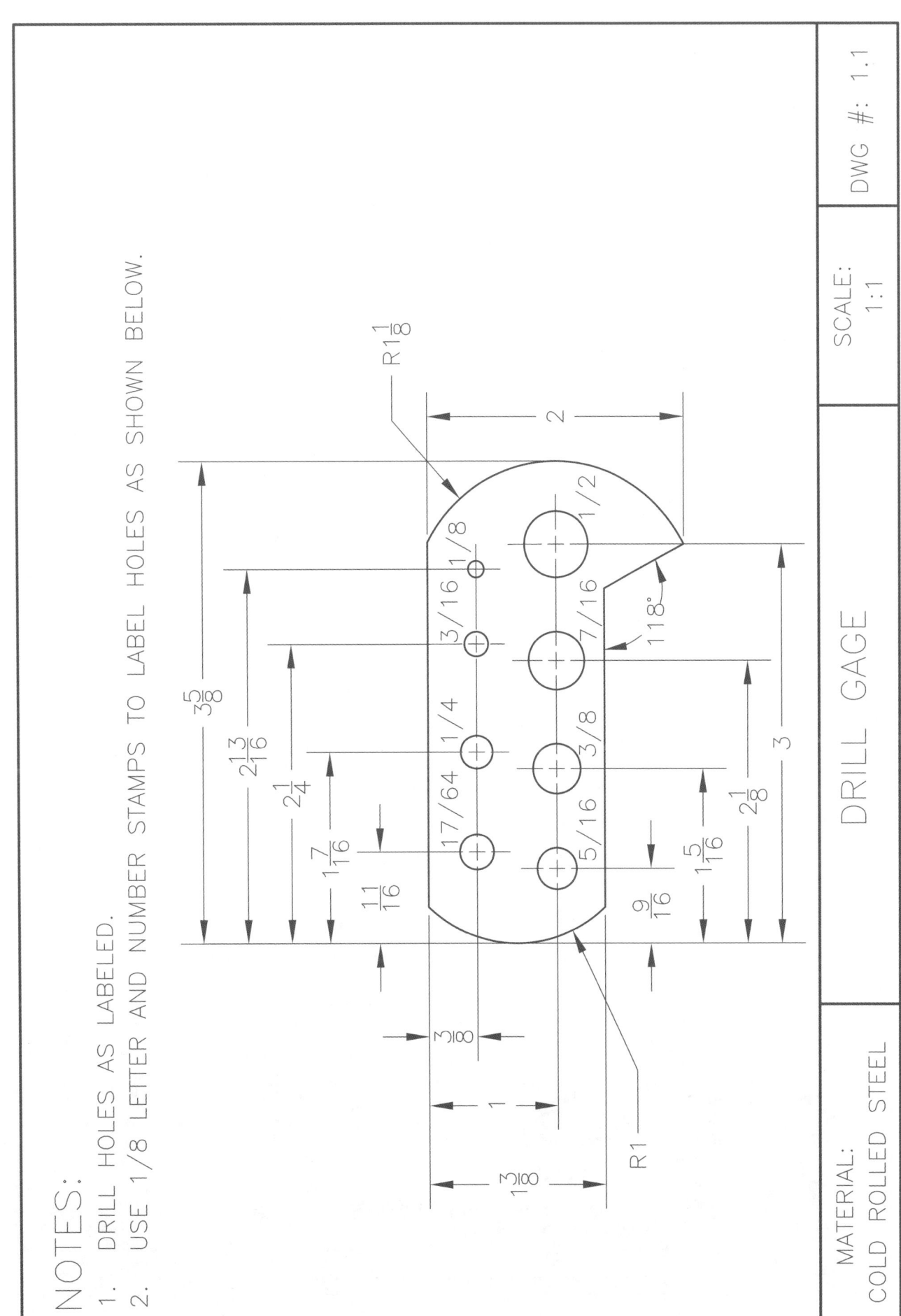

Project 1.2    Drill Drift    *Machining Projects*

# Project 1.2
# Drill Drift

Name _____    Date _____

Instructor _____    Period _____

## NIMS Duties

*The tasks in this project develop skills related to the following NIMS duties:*

- Duty 1.1    Job Process Planning
- Duty 2.1    Manual Operations: Benchwork
- Duty 2.2    Manual Operations: Layout
- Duty 2.8    Drill Press

## Order of Operations

| | |
|---|---|
| 4.1, 12.1–12.6 ❑ | 1. Measure and cut stock plus finishing allowance (approximately 1/8″). |
| 5.2, 5.4, 6.8 ❑ | 2. Square one end with one edge in preparation for layout. |
| 5.1–5.4 ❑ | 3. Perform layout. |
| 10.6 ❑ | 4. Calculate correct cutting speed for the drill bit. |
| 10.9, 10.10 ❑ | 5. Drill and countersink hole. |
| 12.3, 12.4, 12.6 ❑ | 6. Rough cut with vertical band saw. |
| 6.8 ❑ | 7. File the 1/8″ radius along the entire lower edge. |
| 6.8 ❑ | 8. File remaining edges to print specifications. |

## Notes

_____
_____
_____
_____
_____
_____
_____
_____
_____
_____
_____
_____
_____

**Performance Evaluation—Instructor's Use Only**

Project completed on time?    ❑ Yes    ❑ No
    If No, which steps were not completed _____
Overall performance on project:
    ❑ Excellent    ❑ Good    ❑ Satisfactory    ❑ Unsatisfactory    ❑ Poor
Comments: _____
_____

Instructor's Signature _____

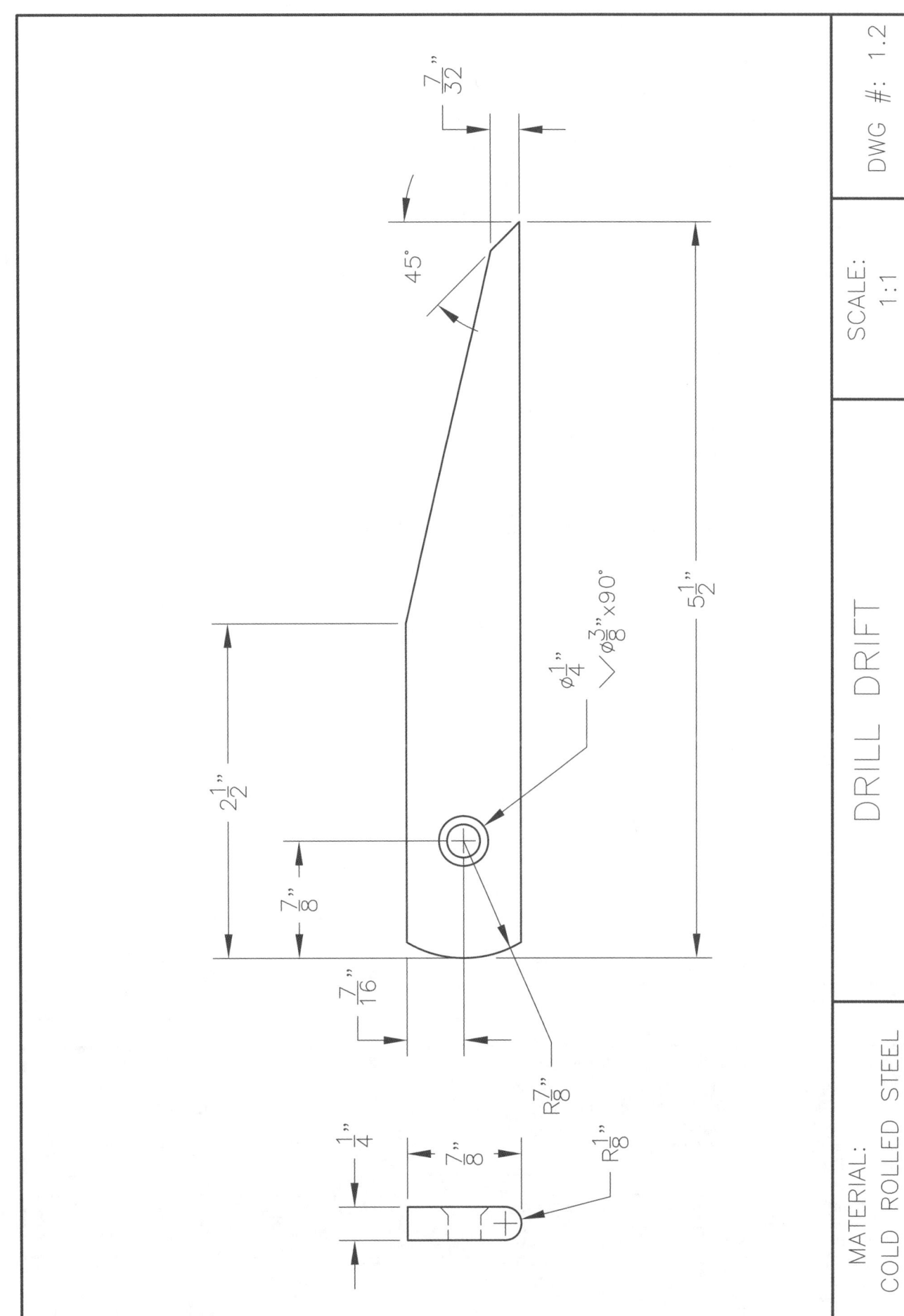

Project 1.3   Drill Index Block                    *Machining Projects*

# Project 1.3
# Drill Index Block

Name _____   Date _____

Instructor _____   Period _____

## NIMS Duties
*The tasks in this project develop skills related to the following NIMS duties:*
- Duty 1.1   Job Process Planning
- Duty 2.1   Manual Operations: Benchwork
- Duty 2.2   Manual Operations: Layout
- Duty 2.8   Drill Press

## Order of Operations

| | |
|---|---|
| 4.1, 12.1–12.6 ❑ | 1. Measure and cut stock plus facing allowance. |
| 5.2, 6.8 ❑ | 2. Square the workpiece to the dimensions specified on the print. |
| 5.1–5.3 ❑ | 3. Perform layout. |
| 10.6 ❑ | 4. Calculate correct cutting speeds for the sizes of drill bits to be used. |
| 10.9 ❑ | 5. Perform drilling operations. |
| ❑ | 6. Stamp drill sizes as indicated on the print. |

## Notes

_____
_____
_____
_____
_____
_____
_____
_____
_____
_____
_____
_____

---

**Performance Evaluation—Instructor's Use Only**

Project completed on time?   ❑ Yes   ❑ No
    If No, which steps were not completed _____

Overall performance on project:
    ❑ Excellent   ❑ Good   ❑ Satisfactory   ❑ Unsatisfactory   ❑ Poor

Comments: _____
_____

Instructor's Signature _____

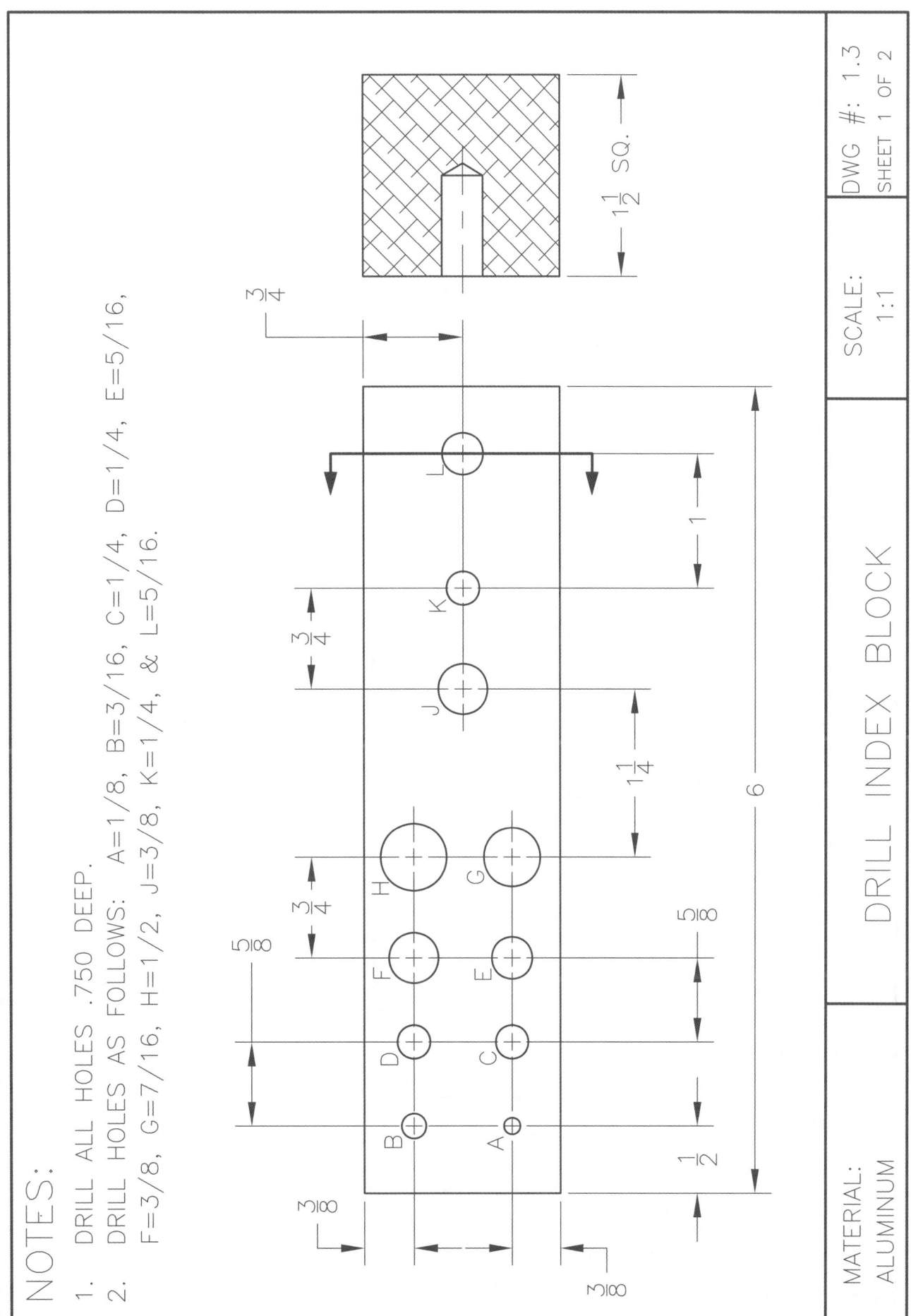

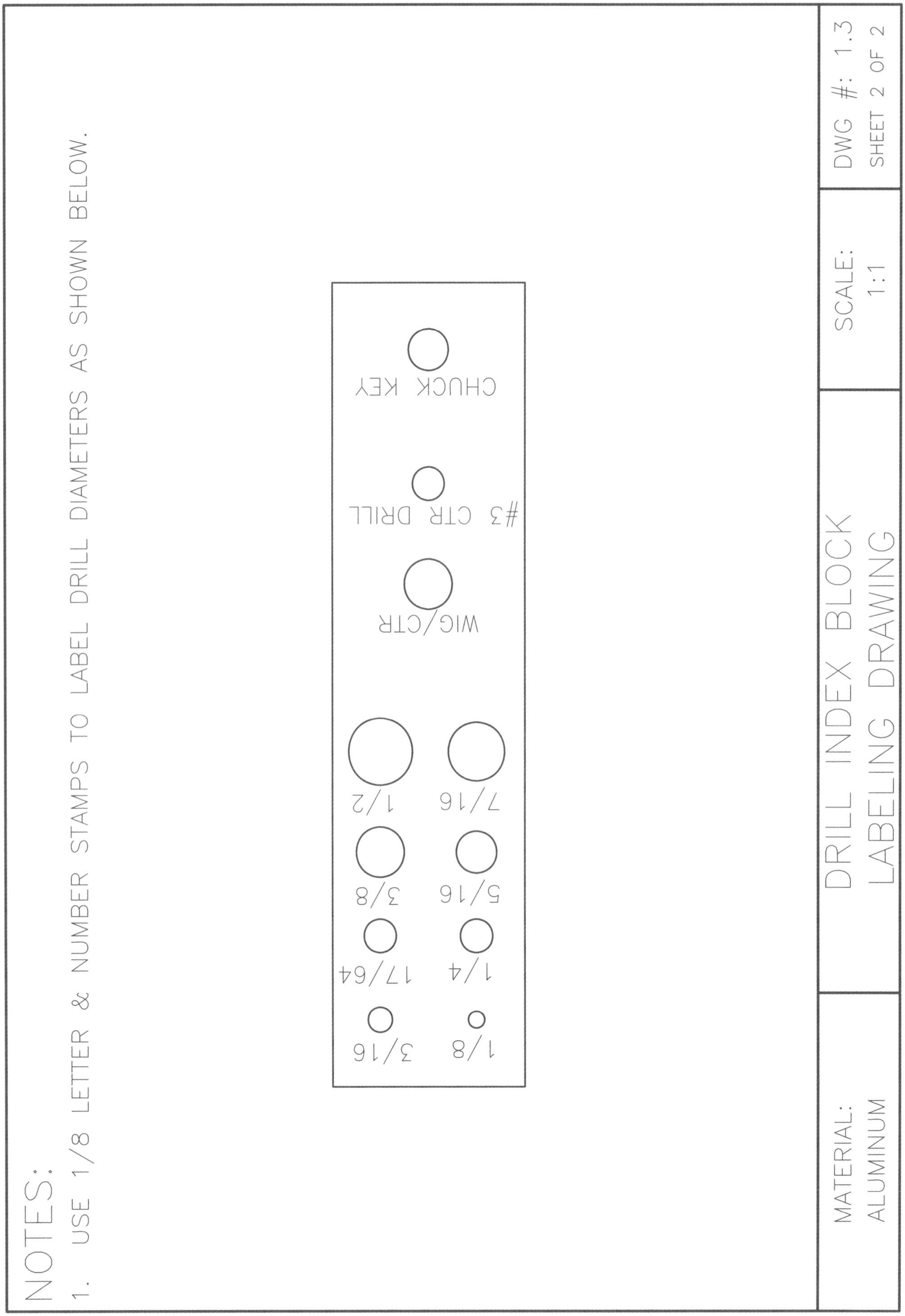

**Project 1.3**  Drill Index Block  *Machining Projects*

NOTES:
1. USE 1/8 LETTER & NUMBER STAMPS TO LABEL DRILL DIAMETERS AS SHOWN BELOW.

Drill hole labels (as shown on block):
- CHUCK KEY
- #3 CTR DRILL
- WIG/CTR
- 1/2, 7/16
- 3/8, 5/16
- 17/64, 1/4
- 3/16, 1/8

DWG #: 1.3
SHEET 2 OF 2
SCALE: 1:1
DRILL INDEX BLOCK
LABELING DRAWING
MATERIAL: ALUMINUM

*Machining Projects*        **Project 1.4**     Drill Index Block

# Project 1.4
# Drill Index Block

Name _____ Date _____

Instructor _____ Period _____

## NIMS Duties

*The tasks in this project develop skills related to the following NIMS duties:*

- ○ Duty 1.1   Job Process Planning
- ○ Duty 2.1   Manual Operations: Benchwork
- ○ Duty 2.2   Manual Operations: Layout
- ○ Duty 2.8   Drill Press

## Order of Operations

**4.1, 12.1–12.6** ❑   1. Measure and cut stock plus finishing allowance.

**5.2, 6.8** ❑   2. Square the workpiece to the specified dimensions.

**5.1–5.4** ❑   3. Perform layout.

**10.6** ❑   4. Calculate correct cutting speeds for the sizes of drill bits to be used.

**10.9** ❑   5. Perform drilling operations.

❑   6. Stamp drill sizes as indicated on the print.

## Notes

_____
_____
_____
_____
_____
_____
_____
_____
_____
_____
_____
_____

---

**Performance Evaluation—Instructor's Use Only**

Project completed on time?   ❑ Yes   ❑ No

     If No, which steps were not completed _____

Overall performance on project:

     ❑ Excellent    ❑ Good    ❑ Satisfactory    ❑ Unsatisfactory    ❑ Poor

Comments: _____

_____

Instructor's Signature _____

**Project 1.4**  Drill Index Block  *Machining Projects*

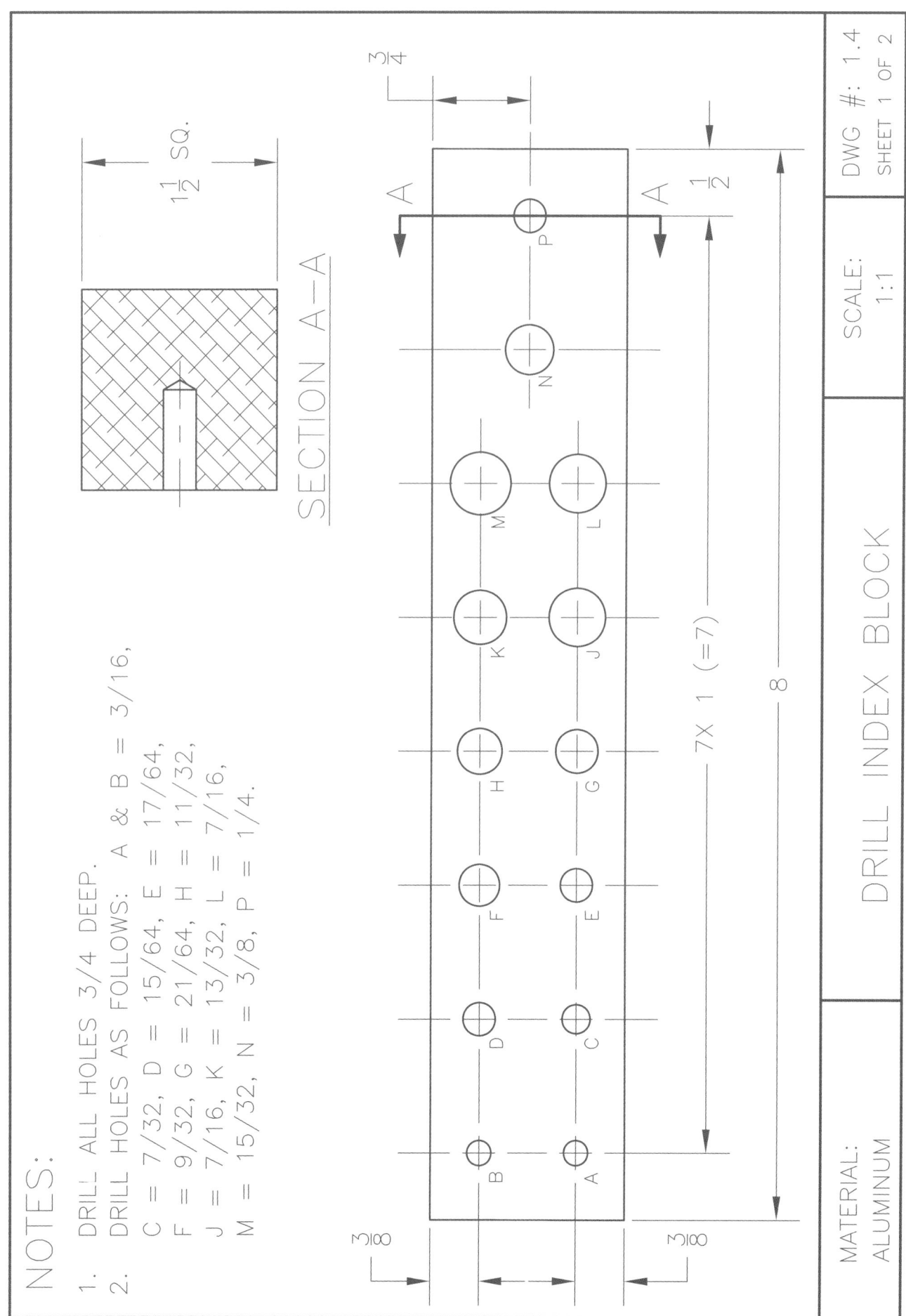

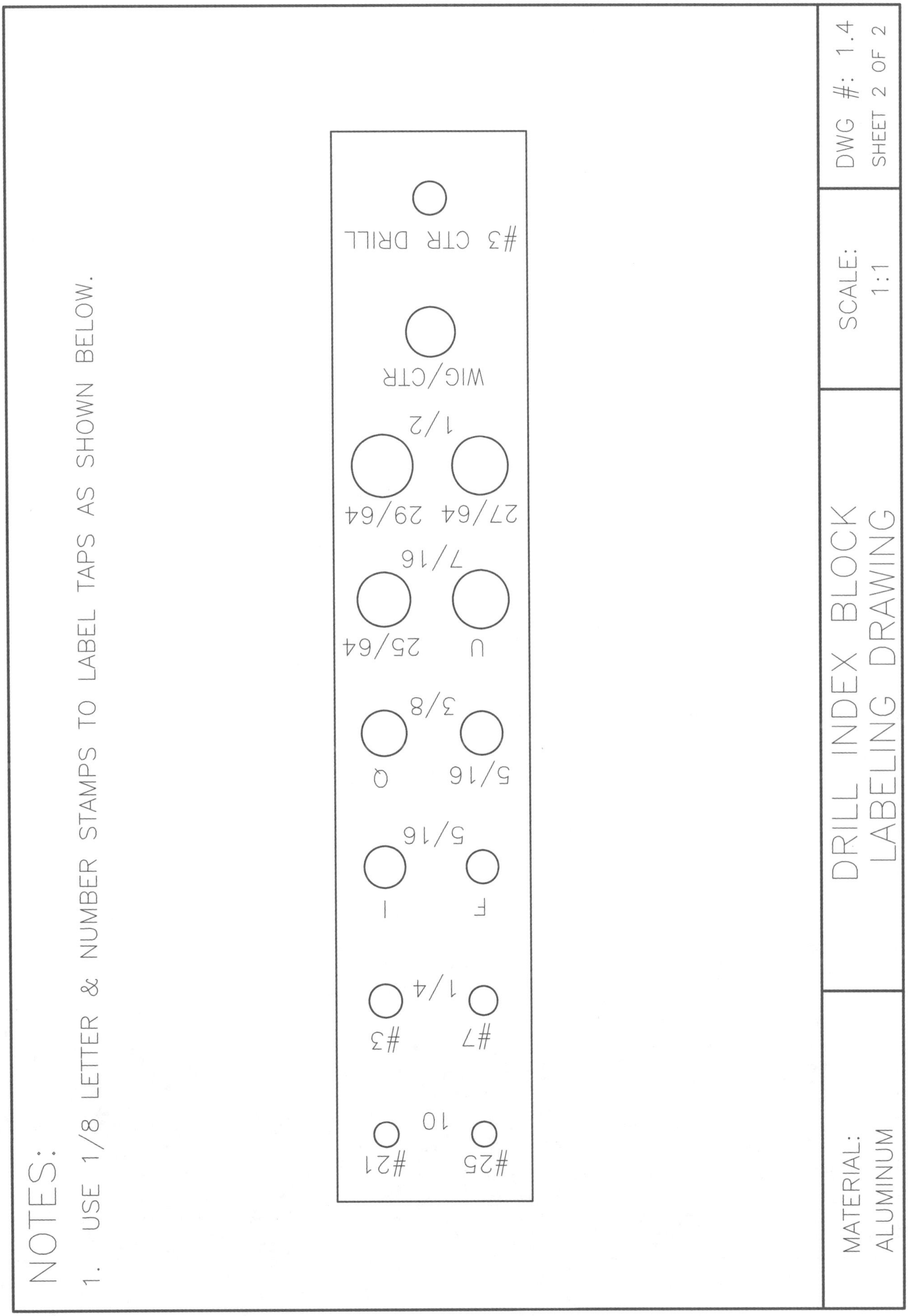

*Machining Projects*      **Project 1.4**    Drill Index Block

**NOTES:**
1. USE 1/8 LETTER & NUMBER STAMPS TO LABEL TAPS AS SHOWN BELOW.

DWG #: 1.4   SHEET 2 OF 2

SCALE: 1:1

DRILL INDEX BLOCK
LABELING DRAWING

MATERIAL: ALUMINUM

Project 1.5   Tap Index Block                                          *Machining Projects*

# Project 1.5
# Tap Index Block

Name _____   Date _____

Instructor _____   Period _____

## NIMS Duties

*The tasks in this project develop skills related to the following NIMS duties:*

○ Duty 1.1   Job Process Planning          ○ Duty 2.2   Manual Operations: Layout
○ Duty 2.1   Manual Operations: Benchwork  ○ Duty 2.8   Drill Press

## Order of Operations

| | | |
|---|---|---|
| 4.1, 12.1–12.6 | ❑ | 1. Measure and cut stock plus finishing allowance. |
| 5.2, 6.8 | ❑ | 2. Square workpiece to specifications given on the print. |
| 5.1–5.4 | ❑ | 3. Perform layout. |
| 10.6 | ❑ | 4. Calculate the correct cutting speeds for the sizes of drill bits to be used. |
| 10.9 | ❑ | 5. Perform drilling operations. |
|  | ❑ | 6. Stamp thread sizes as indicated on the print. |

## Notes

_____
_____
_____
_____
_____
_____
_____
_____
_____
_____
_____
_____
_____
_____

---

**Performance Evaluation—Instructor's Use Only**

Project completed on time?    ❑ Yes    ❑ No
    If No, which steps were not completed _____
Overall performance on project:
    ❑ Excellent    ❑ Good    ❑ Satisfactory    ❑ Unsatisfactory    ❑ Poor
Comments: _____
_____

Instructor's Signature_____

Machining Projects  Project 1.5  Tap Index Block

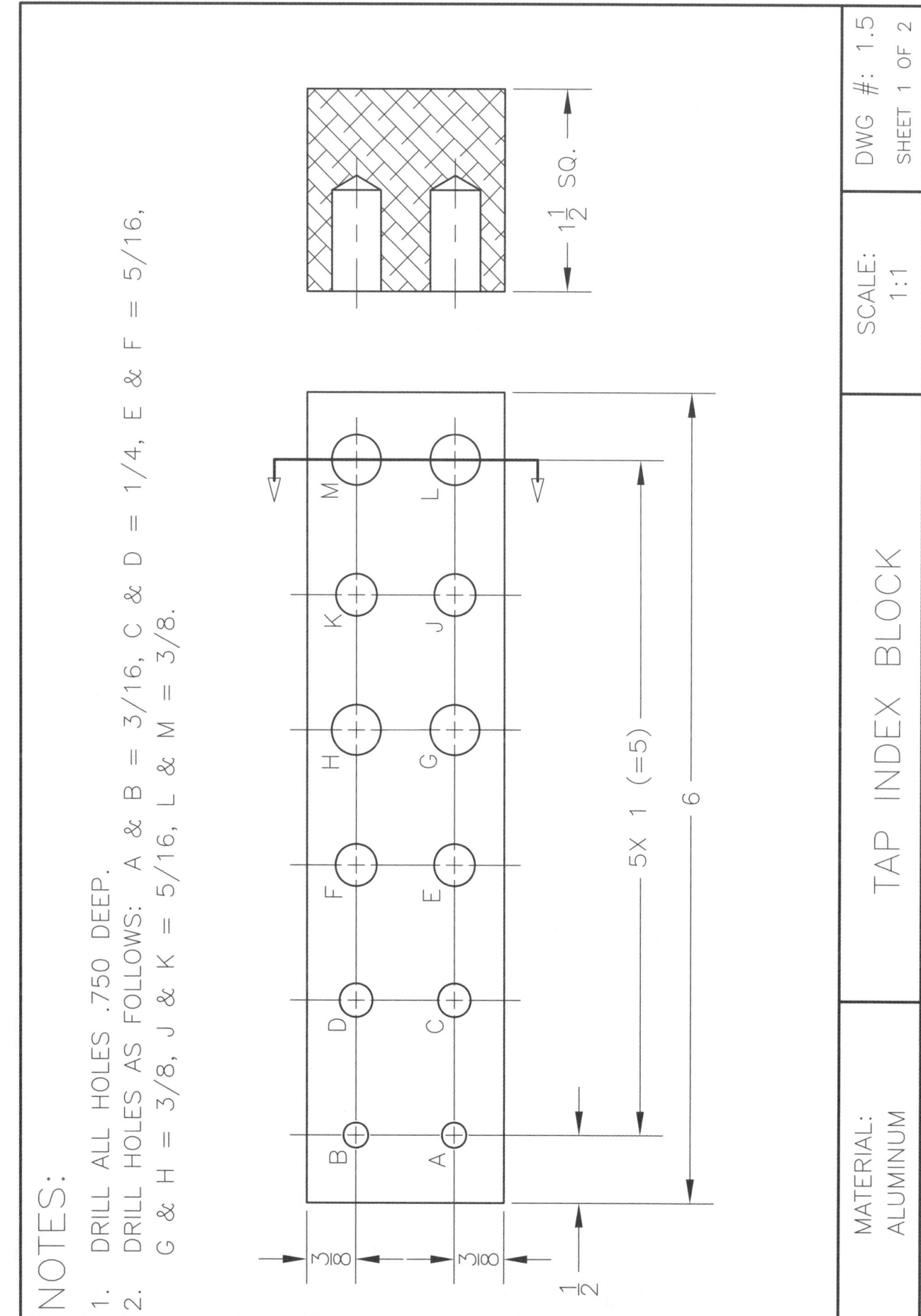

NOTES:
1. DRILL ALL HOLES .750 DEEP.
2. DRILL HOLES AS FOLLOWS: A & B = 3/16, C & D = 1/4, E & F = 5/16, G & H = 3/8, J & K = 5/16, L & M = 3/8.

TAP INDEX BLOCK

MATERIAL: ALUMINUM
SCALE: 1:1
DWG #: 1.5
SHEET 1 OF 2

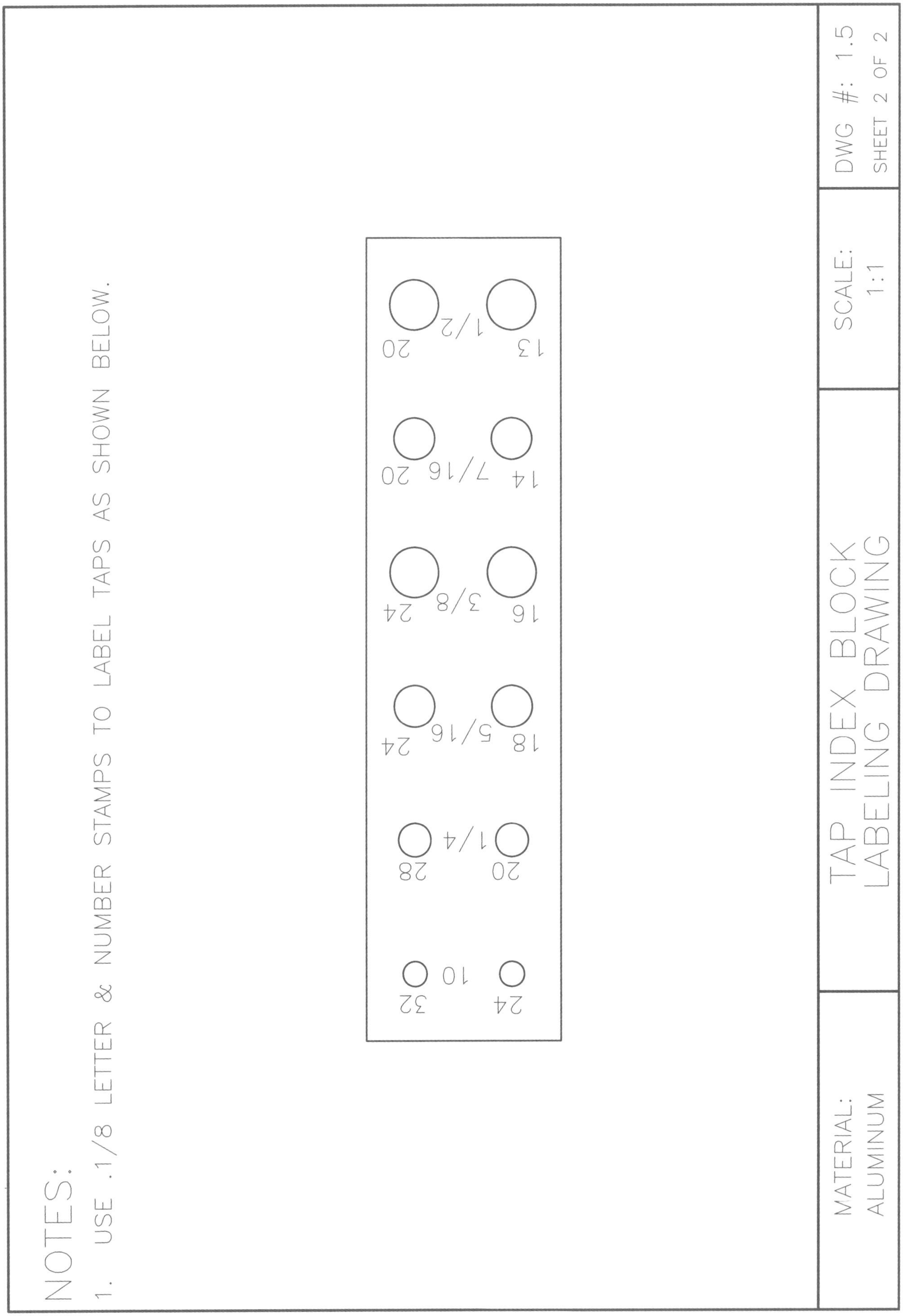

**Project 1.5**  Tap Index Block  *Machining Projects*

NOTES:
1. USE .1/8 LETTER & NUMBER STAMPS TO LABEL TAPS AS SHOWN BELOW.

Tap labels (as shown on the block):
- 24   10/32
- 20  1/4  28
- 18  5/16  24
- 16  3/8  24
- 14  7/16  20
- 13  1/2  20

DWG #: 1.5
SHEET 2 OF 2
SCALE: 1:1
TAP INDEX BLOCK
LABELING DRAWING
MATERIAL: ALUMINUM

# Project 1.6
# Tapping Plate

Name _____ Date _____

Instructor _____ Period _____

## NIMS Duties

*The tasks in this project develop skills related to the following NIMS duties:*
- Duty 1.1   Job Process Planning
- Duty 2.1   Manual Operations: Benchwork
- Duty 2.2   Manual Operations: Layout
- Duty 2.8   Drill Press

## Order of Operations

| | | |
|---:|:---:|---|
| 4.1, 12.1–12.6 | ❑ | 1. Measure and cut stock plus finishing allowance. |
| 5.2, 6.8 | ❑ | 2. Square the workpiece to specified dimensions. |
| 5.1–5.4 | ❑ | 3. Perform layout. |
| 10.6 | ❑ | 4. Calculate the correct cutting speeds for the sizes of drill bits to be used. |
| 10.9, 10.10 | ❑ | 5. Drill and countersink all holes. |
| 6.10 | ❑ | 6. Tap all holes with the correct taps. |
| | ❑ | 7. Using letter and number stamps, stamp all illustrated information as it appears on the print. |

## Notes

_____
_____
_____
_____
_____
_____
_____
_____
_____
_____
_____
_____
_____

---

**Performance Evaluation—Instructor's Use Only**

Project completed on time?    ❑ Yes    ❑ No
    If No, which steps were not completed _____
Overall performance on project:
    ❑ Excellent    ❑ Good    ❑ Satisfactory    ❑ Unsatisfactory    ❑ Poor
Comments: _____
_____

Instructor's Signature _____

# Project 1.6 Tapping Plate

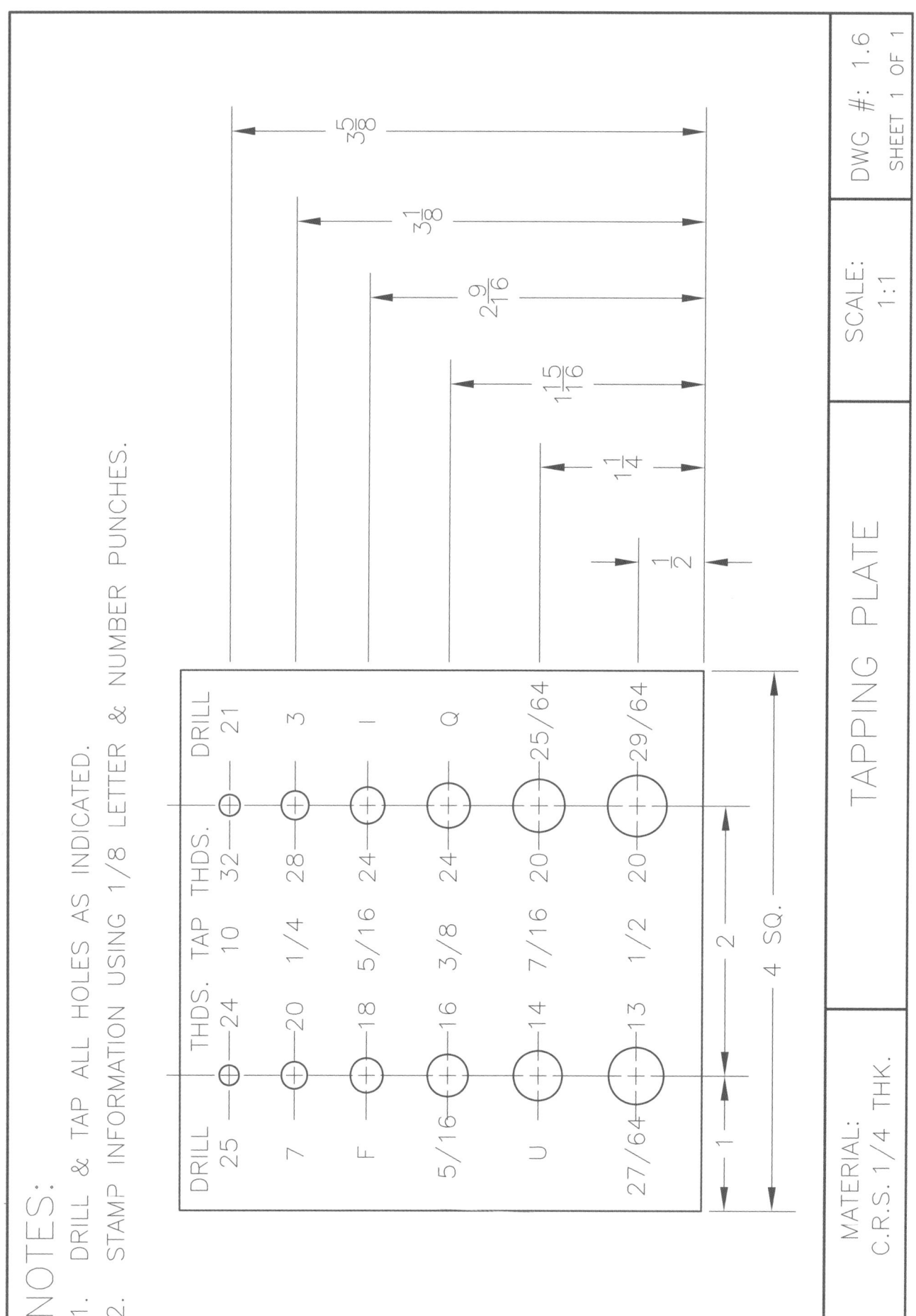

Machining Projects

# Section 2
## Lathe Competency

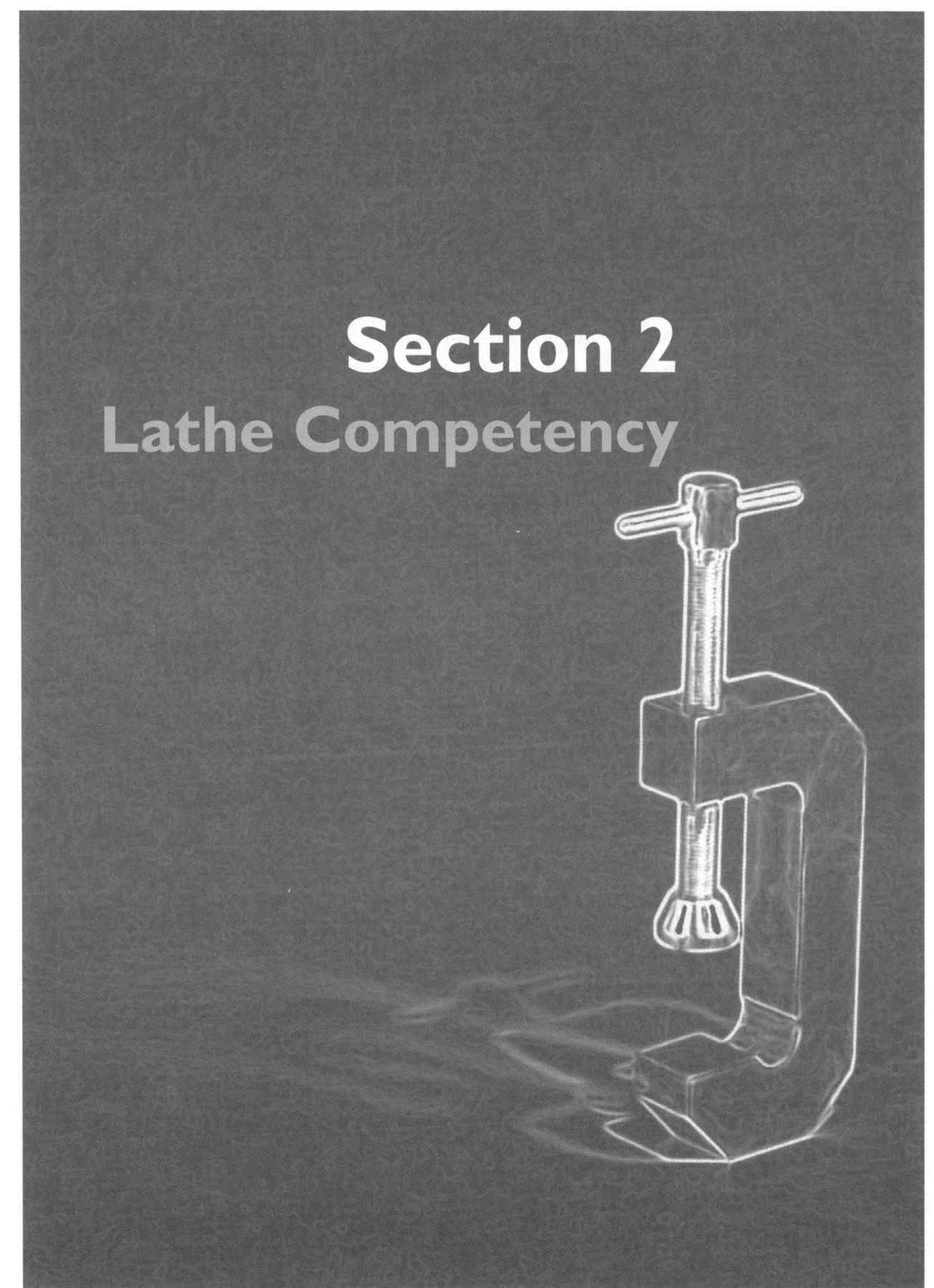

# Project 2.1
# Faced Bar

Name _____ Date _____

Instructor _____ Period _____

## NIMS Duties

*The tasks in this project develop skills related to the following NIMS duties:*

- ○ Duty 1.1    Job Process Planning
- ○ Duty 2.4    Turning Operations: Chucking
- ○ Duty 2.7a   Grinding Wheel Safety

## Order of Operations

| | | |
|---|---|---|
| **4.1, 12.1–12.6** | ❑ | 1. Select, measure, and cut appropriate stock plus facing allowance. |
| **13.6** | ❑ | 2. Select or grind appropriate lathe tool bit. |
| **13.11** | ❑ | 3. Set up lathe for facing operation. |
| **13.7** | ❑ | 4. Calculate and set appropriate speed and feed. |
| **13.2.3, 13.11** | ❑ | 5. Set feed direction. |
| **13.11** | ❑ | 6. Face the workpiece to the specified length. |

## Notes

_____
_____
_____
_____
_____
_____
_____
_____
_____
_____
_____
_____
_____

---

**Performance Evaluation—Instructor's Use Only**

Project completed on time? ❑ Yes ❑ No

    If No, which steps were not completed _____

Overall performance on project:

    ❑ Excellent    ❑ Good    ❑ Satisfactory    ❑ Unsatisfactory    ❑ Poor

Comments: _____
_____

Instructor's Signature _____

*Machining Projects*        **Project 2.1**     Faced Bar

NOTES:
1. BREAK SHARP EDGES.

- 2.000
- Ø.625

DWG #: 2.1
SCALE: 1:1
FACED BAR
MATERIAL: COLD ROLLED STEEL

# Project 2.3
# Step Bar

Name _____  Date _____

Instructor _____  Period _____

## NIMS Duties

*The tasks in this project develop skills related to the following NIMS duties:*
- Duty 1.1  Job Process Planning
- Duty 2.3  Turning Operations: Between Centers Turning
- Duty 2.4  Turning Operations: Chucking
- Duty 2.7a  Grinding Wheel Safety

## Order of Operations

| | | |
|---|---|---|
| 4.1, 12.1–12.6 | ❑ | 1. Measure and cut stock plus facing allowance. |
| 13.11 | ❑ | 2. Face to specified length. |
| 13.9, 15.2 | ❑ | 3. Center drill. |
| 13.9–13.12 | ❑ | 4. Straight turn the workpiece to the specified lengths and diameters. |

## Notes

_____

_____

_____

_____

_____

_____

_____

_____

_____

_____

_____

_____

_____

_____

_____

---

**Performance Evaluation—Instructor's Use Only**

Project completed on time?  ❑ Yes  ❑ No
   If No, which steps were not completed _____

Overall performance on project:
   ❑ Excellent  ❑ Good  ❑ Satisfactory  ❑ Unsatisfactory  ❑ Poor

Comments: _____

_____

Instructor's Signature _____

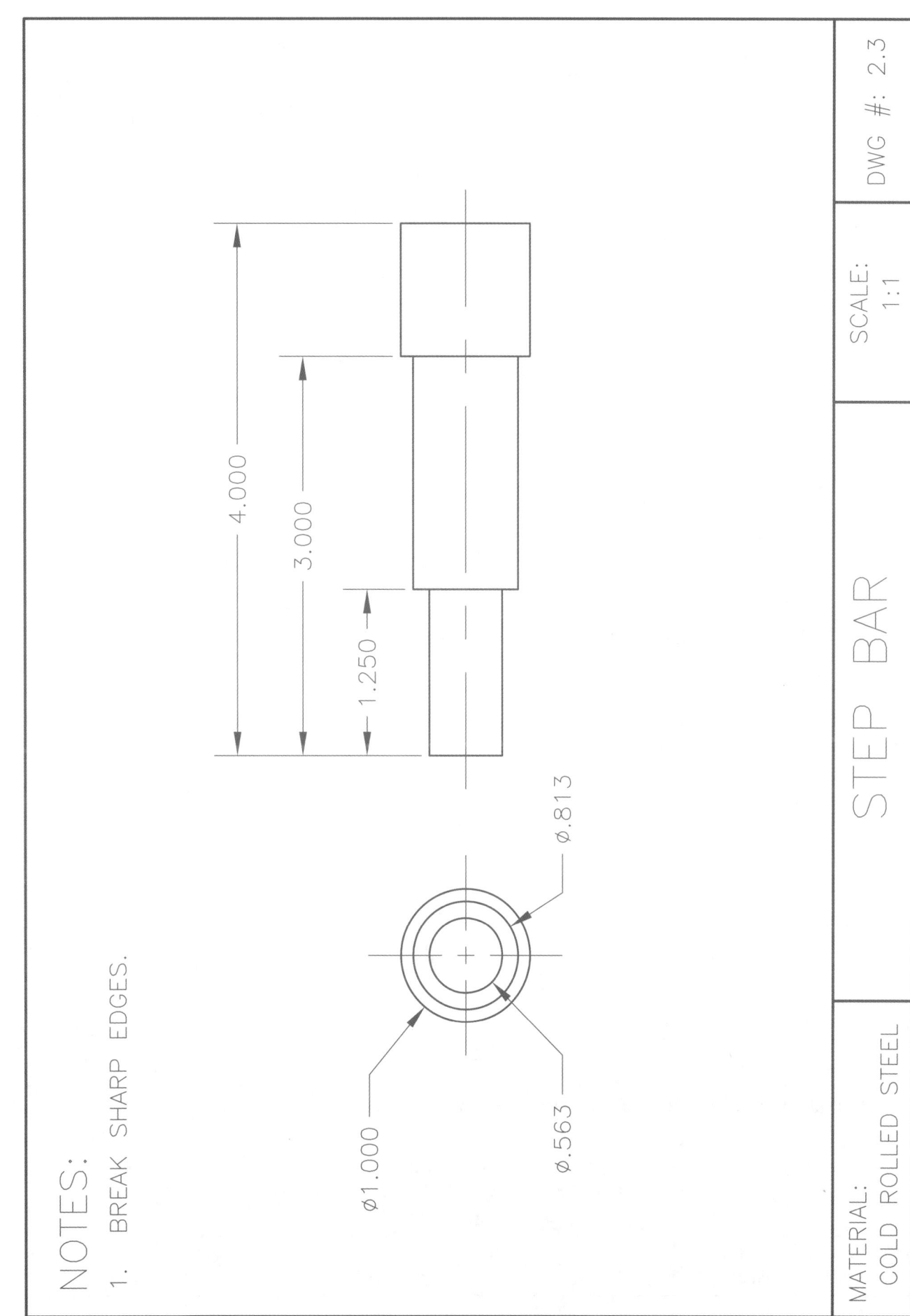

# Project 2.4.1
# Threaded Bar

Name _____ Date _____

Instructor _____ Period _____

## NIMS Duties

*The tasks in this project develop skills related to the following NIMS duties:*
- Duty 1.1  Job Process Planning
- Duty 2.3  Turning Operations: Between Centers Turning
- Duty 2.4  Turning Operations: Chucking
- Duty 2.7a Grinding Wheel Safety

## Order of Operations

| | | |
|---|---|---|
| 4.1, 12.1–12.6 | ☐ | 1. Select, measure, and cut stock plus facing allowance. |
| 13.11 | ☐ | 2. Set up and face to specified length. |
| 13.9.1, 15.2 | ☐ | 3. Center drill both ends. |
| 13.9–13.12 | ☐ | 4. Straight turn ends to specified lengths and diameters. |
| 6.10, 6.14 | ☐ | 5. Chamfer end of workpiece. |
| 6.10 | ☐ | 6. Assemble die and die stock. |
| 6.10 | ☐ | 7. Set up tailstock spindle for support of the die and die stock. |

Note   The face of a Jacobs chuck mounted in the tailstock spindle can also be used for this purpose.

| | | |
|---|---|---|
| 6.10 | ☐ | 8. Using correct technique, cut threads as indicated. |
| 6.10 | ☐ | 9. Reverse the die to cut threads to the shoulder of the workpiece. |

## Notes

_____
_____
_____
_____
_____
_____
_____
_____
_____
_____

---

**Performance Evaluation—Instructor's Use Only**

Project completed on time?   ☐ Yes   ☐ No
   If No, which steps were not completed _____

Overall performance on project:
   ☐ Excellent   ☐ Good   ☐ Satisfactory   ☐ Unsatisfactory   ☐ Poor

Comments: _____
_____

Instructor's Signature _____

Machining Projects — Project 2.4.1 — Threaded Bar

NOTES:
1. BREAK SHARP EDGES.

- $2\frac{1}{2}$
- $1\frac{3}{4}$
- 1/2–13UNC–2A
- 1/8 × 45° CHAMFER
- $\varnothing \frac{5}{8}$
- $\varnothing \frac{1}{2}$

THREADED BAR

DWG #: 2.4.1
SCALE: 1:1
MATERIAL: COLD ROLLED STEEL

# Project 2.4.2
# Threaded Nut

Name _____ Date _____

Instructor _____ Period _____

## NIMS Duties

*The tasks in this project develop skills related to the following NIMS duties:*
- Duty 1.1  Job Process Planning
- Duty 2.4  Turning Operations: Chucking
- Duty 2.7a  Grinding Wheel Safety

## Order of Operations

| | | |
|---|---|---|
| 4.1, 12.1–12.6 | ❏ | 1. Select, measure, and cut stock to length plus facing allowance. |
| 13.7 | ❏ | 2. Calculate appropriate speed and feed and adjust lathe accordingly. |
| 13.11 | ❏ | 3. Set up lathe and face workpiece to the specified length. |
| 13.9, 15.2 | ❏ | 4. Center drill one end of the workpiece. |
| 6.10 | ❏ | 5. Using a decimal equivalent and tap drill chart as a reference, locate the correct tap drill for a 1/2-13 thread and drill the workpiece through. |
| 6.10 | ❏ | 6. Countersink both ends of the workpiece to the major diameter of the thread. |
| 6.10 | ❏ | 7. Place a 1/2-13 tap in a tap wrench, support it with a live center mounted in the tailstock of the lathe, brush on some cutting oil, and advance the tap until full diameter threads have been cut completely through to nut. |

## Notes

_____
_____
_____
_____
_____
_____
_____
_____
_____
_____
_____

---

**Performance Evaluation—Instructor's Use Only**
Project completed on time?   ❏ Yes   ❏ No
  If No, which steps were not completed _____
Overall performance on project:
  ❏ Excellent   ❏ Good   ❏ Satisfactory   ❏ Unsatisfactory   ❏ Poor
Comments: _____

Instructor's Signature _____

Machining Projects — Project 2.4.2 — Threaded Nut

- 1/16 × 45° CHAMFER BOTH SIDES
- 5/8
- 1/2–13UNC–2B
- 1

DWG #: 2.4.2
SCALE: 1:1
THREADED NUT
MATERIAL: COLD ROLLED STEEL

# Project 2.5.1
# Turned Bar—Undercut and Chased Threads

Name _____ Date _____

Instructor _____ Period _____

## NIMS Duties
*The tasks in this project develop skills related to the following NIMS duties:*
- Duty 1.1  Job Process Planning
- Duty 2.3  Turning Operations: Between Centers Turning
- Duty 2.4  Turning Operations: Chucking
- Duty 2.7a Grinding Wheel Safety

## Order of Operations

| | |
|---|---|
| 4.1, 12.1–12.6 ❏ | 1. Measure and cut stock plus facing allowance. |
| 13.11 ❏ | 2. Face to specified length. |
| 13.9, 15.2 ❏ | 3. Center drill. |
| 13.9, 15.2 ❏ | 4. Straight turn section to be threaded to specified length and diameter. |
| 6.10, 14.6 ❏ | 5. Cut chamfer. |
| 14.6 ❏ | 6. Cut relief groove for the termination of the threads. |
| 14.6 ❏ | 7. Set up the lathe for chasing threads. |
| 14.6 ❏ | 8. Chase threads to specifications. |

## Notes

_____
_____
_____
_____
_____
_____
_____
_____
_____
_____
_____
_____

**Performance Evaluation—Instructor's Use Only**
Project completed on time?   ❏ Yes   ❏ No
  If No, which steps were not completed _____
Overall performance on project:
  ❏ Excellent   ❏ Good   ❏ Satisfactory   ❏ Unsatisfactory   ❏ Poor
Comments: _____
_____
Instructor's Signature _____

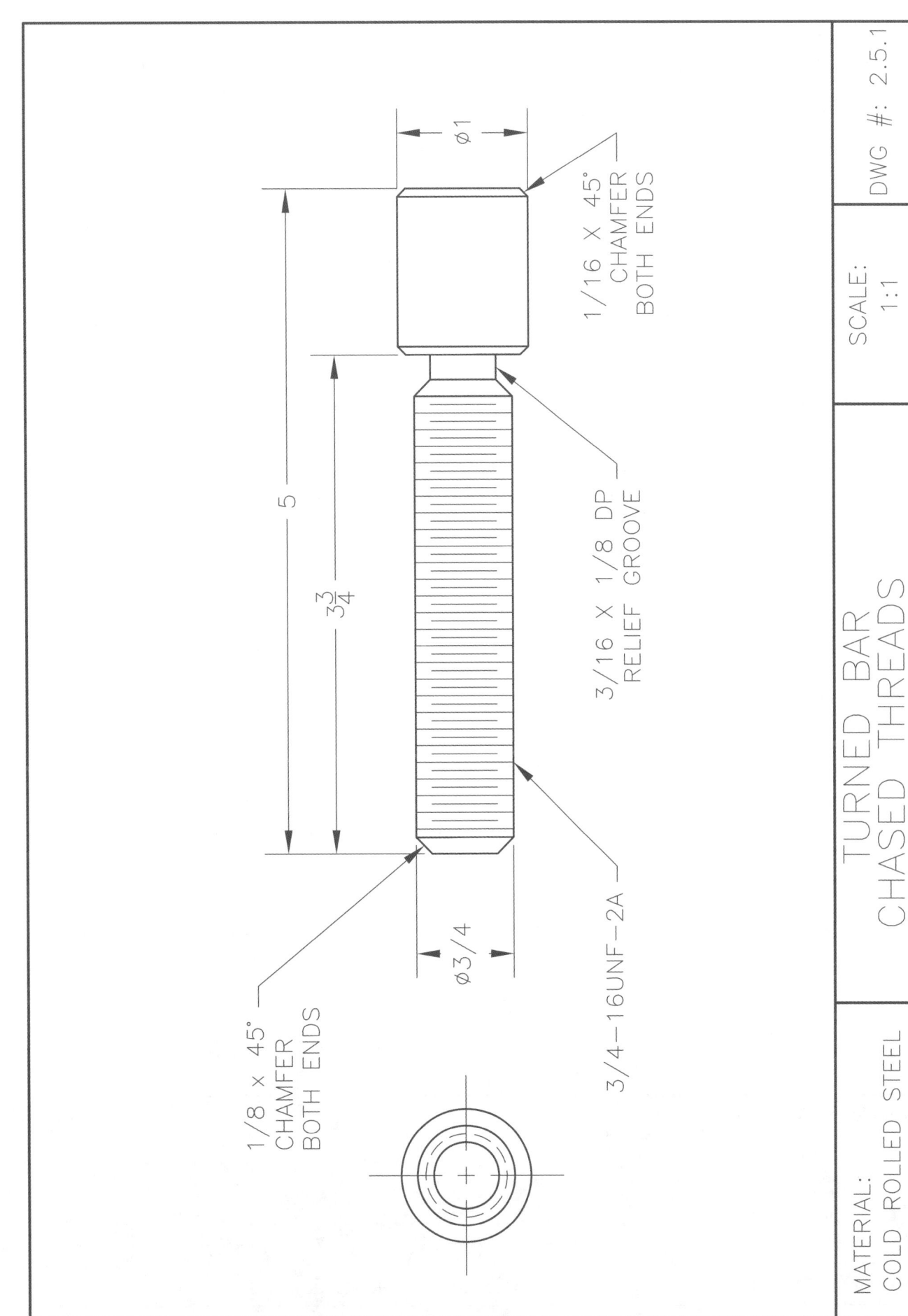

Project 2.5.2  Threaded Nut—Undercut and Chased Threads     Machining Projects

# Project 2.5.2
# Threaded Nut—Undercut and Chased Threads

Name _____  Date _____

Instructor _____  Period _____

## NIMS Duties
*The tasks in this project develop skills related to the following NIMS duties:*
- Duty 1.1  Job Process Planning
- Duty 2.4  Turning Operations: Chucking
- Duty 2.7a  Grinding Wheel Safety

## Order of Operations

| | | |
|---|---|---|
| 4.1, 12.1–12.6 | ❑ | 1. Measure and cut stock plus facing allowance. |
| 13.11 | ❑ | 2. Face workpiece to specifications. |
| 13.9, 15.2 | ❑ | 3. Center drill and drill through using the appropriate drill bit. |
| 10.10 | ❑ | 4. Countersink to major diameter (both sides). |
| 14.6 | ❑ | 5. Set up lathe for chasing interior threads using a boring bar and properly ground tool bit. |
| 14.6 | ❑ | 6. Chase threads to specifications. |

## Notes

_____
_____
_____
_____
_____
_____
_____
_____
_____
_____
_____
_____
_____

---

**Performance Evaluation—Instructor's Use Only**

Project completed on time?   ❑ Yes   ❑ No
    If No, which steps were not completed _____
Overall performance on project:
    ❑ Excellent   ❑ Good   ❑ Satisfactory   ❑ Unsatisfactory   ❑ Poor
Comments: _____
_____

Instructor's Signature _____

Machining Projects  Project 2.5.2  Threaded Nut—Undercut and Chased Threads

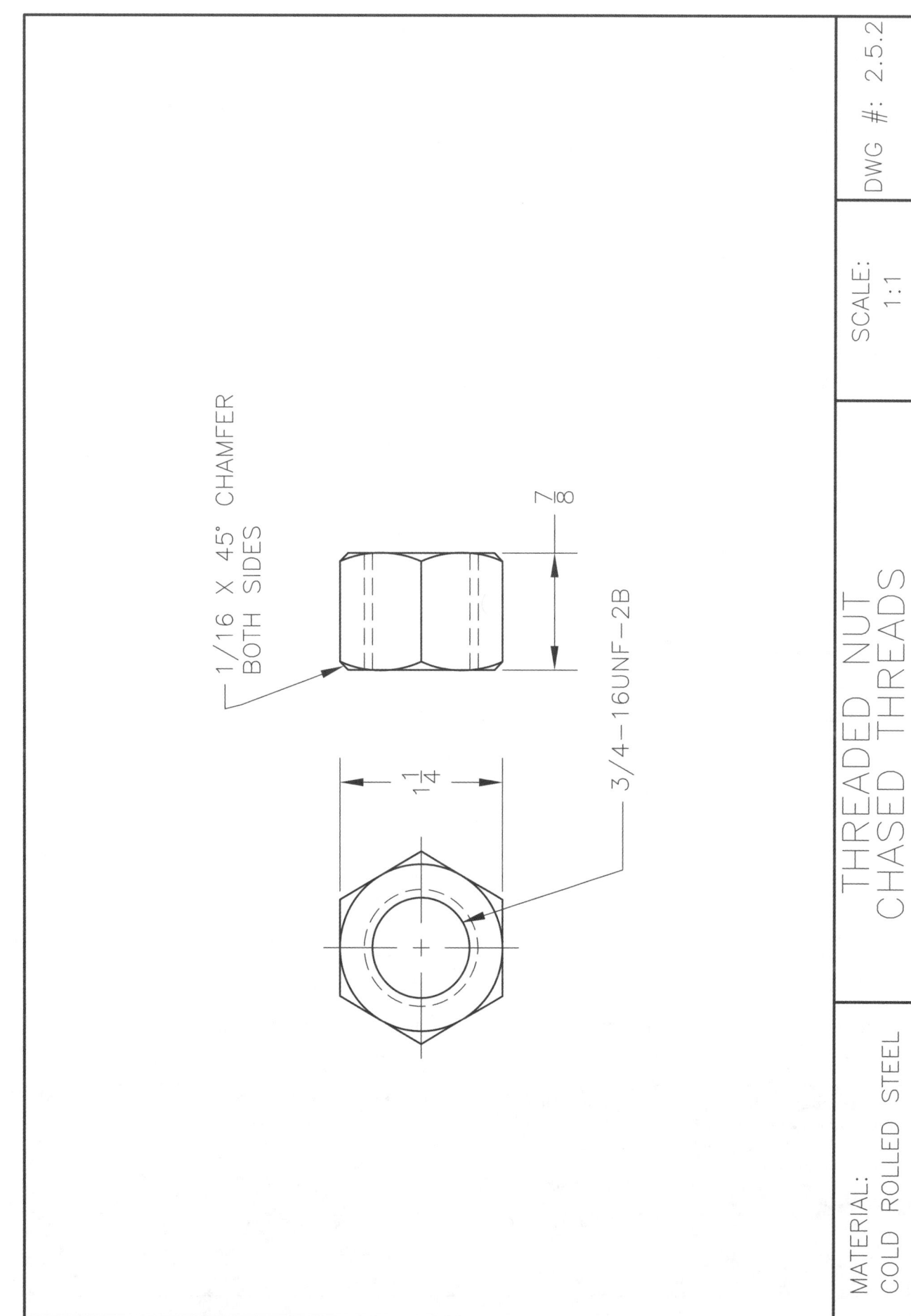

Project 2.6   Knurled Bar                                                          *Machining Projects*

# Project 2.6
# Knurled Bar

Name _____   Date _____

Instructor _____   Period _____

## NIMS Duties
*The tasks in this project develop skills related to the following NIMS duties:*
- Duty 1.1   Job Process Planning
- Duty 2.3   Turning Operations: Between Centers Turning
- Duty 2.4   Turning Operations: Chucking
- Duty 2.7a  Grinding Wheel Safety

## Order of Operations

| | | |
|---|---|---|
| 4.1, 12.1–12.6 | ❑ | 1. Measure and cut stock to specified length plus facing allowance. |
| 13.11 | ❑ | 2. Face to specified length. |
| 13.9, 15.2 | ❑ | 3. Center drill both ends. |
| 15.3 | ❑ | 4. Set up lathe for knurling. |
| 15.3 | ❑ | 5. Perform knurling operation. |
| 13.9–13.12 | ❑ | 6. Turn ends to specified lengths and diameters. |
| 14.1 | ❑ | 7. Set lathe compound and cut chamfers. |

## Notes

_____
_____
_____
_____
_____
_____
_____
_____
_____
_____
_____
_____

---

**Performance Evaluation—Instructor's Use Only**
Project completed on time?   ❑ Yes   ❑ No
   If No, which steps were not completed _____
Overall performance on project:
   ❑ Excellent   ❑ Good   ❑ Satisfactory   ❑ Unsatisfactory   ❑ Poor
Comments: _____
_____

Instructor's Signature _____

Machining Projects  Project 2.6  Knurled Bar

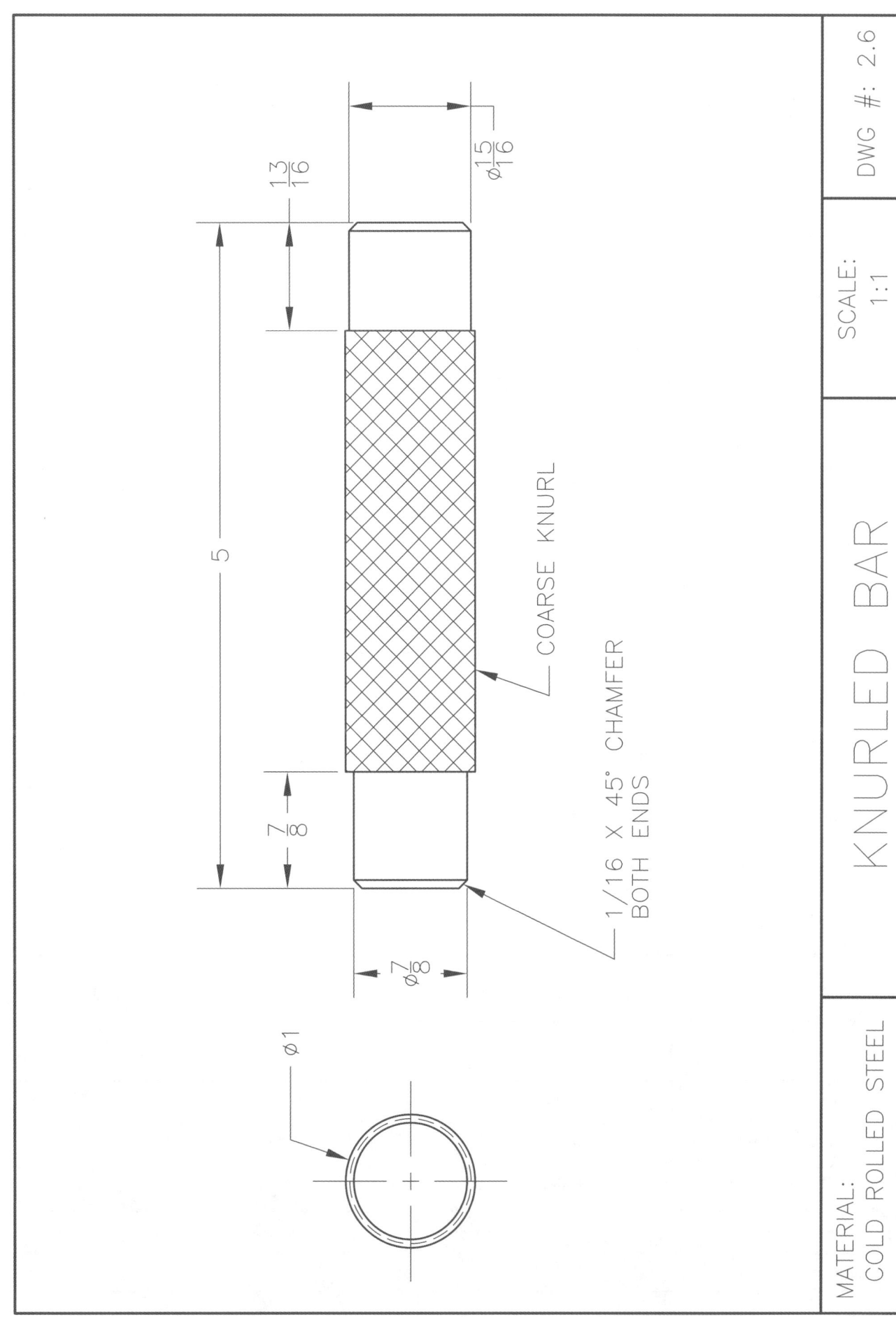

# Project 2.7
# Go, No-Go Bar

Name _____ Date _____

Instructor _____ Period _____

## NIMS Duties
*The tasks in this project develop skills related to the following NIMS duties:*
- Duty 1.1  Job Process Planning
- Duty 2.3  Turning Operations: Between Centers Turning
- Duty 2.4  Turning Operations: Chucking
- Duty 2.7a Grinding Wheel Safety

## Order of Operations

| | | |
|---|---|---|
| 4.1, 12.1–12.6 | ❏ | 1. Measure and cut stock plus facing allowance. |
| 13.11 | ❏ | 2. Face to specified length. |
| 13.9 | ❏ | 3. Center drill each end. |
| 15.3 | ❏ | 4. Perform knurling operations. |
| 13.9–13.12 | ❏ | 5. Straight turn steps to the specified lengths and diameters. |
| 13.9 | ❏ | 6. Perform undercutting operations. |

## Notes

_____
_____
_____
_____
_____
_____
_____
_____
_____
_____
_____
_____
_____

---

**Performance Evaluation—Instructor's Use Only**

Project completed on time?  ❏ Yes  ❏ No

If No, which steps were not completed _____

Overall performance on project:
❏ Excellent  ❏ Good  ❏ Satisfactory  ❏ Unsatisfactory  ❏ Poor

Comments: _____
_____
_____

Instructor's Signature _____

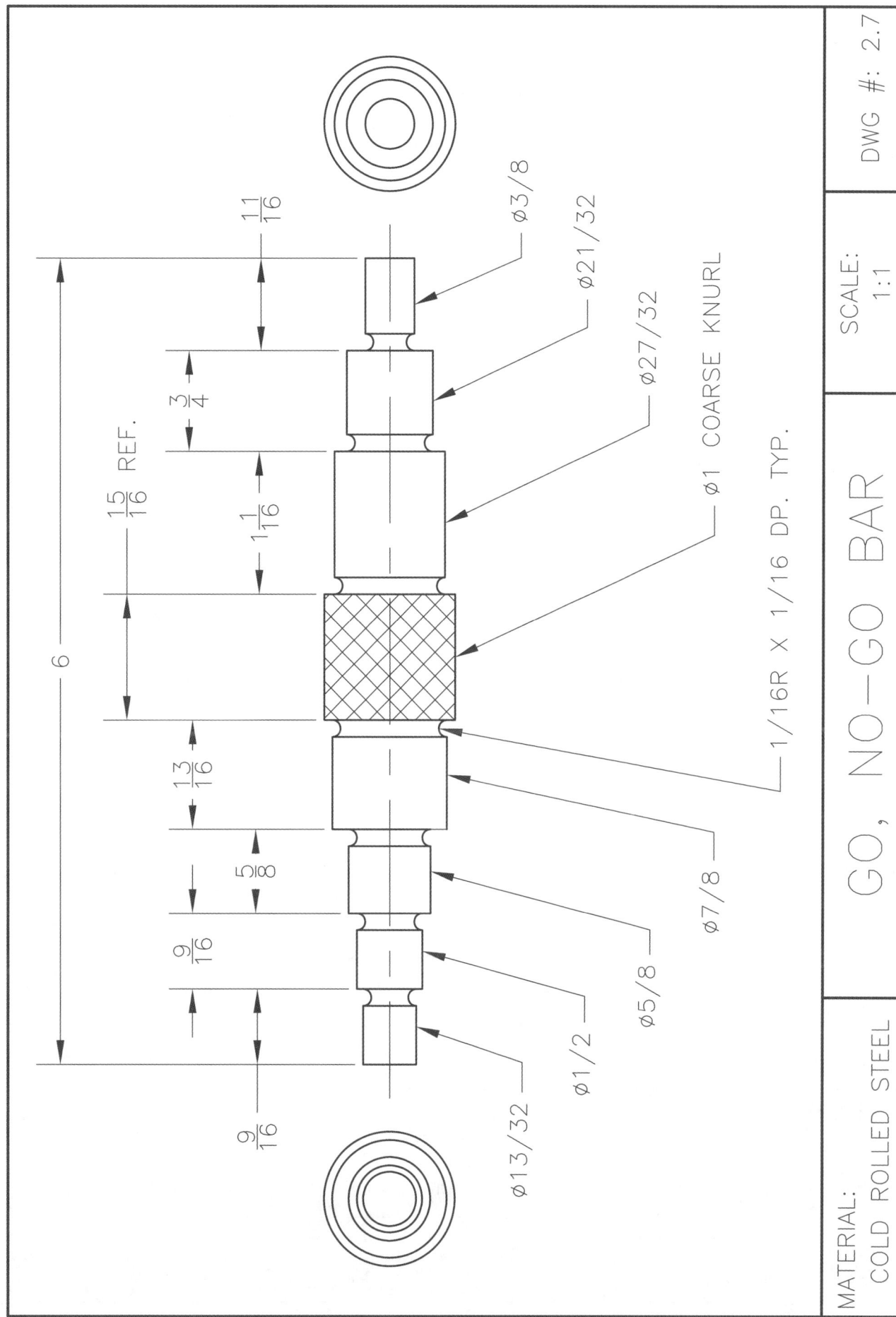

Project 2.8    Bored Bar    *Machining Projects*

# Project 2.8
# Bored Bar

Name _____    Date _____

Instructor _____    Period _____

## NIMS Duties
*The tasks in this project develop skills related to the following NIMS duties:*
- Duty 1.1    Job Process Planning
- Duty 2.4    Turning Operations: Chucking
- Duty 2.7a   Grinding Wheel Safety

## Order of Operations

| | | |
|---|---|---|
| 4.1, 12.1–12.6 | ❏ | 1. Measure and cut stock to length plus facing allowance. |
| 13.11 | ❏ | 2. Face to specified length. |
| 13.9, 15.2 | ❏ | 3. Center drill and drill starter hole through. |
| 15.1–15.2 | ❏ | 4. Set up lathe with a boring bar and bore hole to the specified diameter. |
| 14.1 | ❏ | 5. Set lathe compound and cut chamfers. |

**Notes**

_____
_____
_____
_____
_____
_____
_____
_____
_____
_____
_____
_____
_____

---

**Performance Evaluation—Instructor's Use Only**

Project completed on time?    ❏ Yes    ❏ No
    If No, which steps were not completed _____
Overall performance on project:
    ❏ Excellent    ❏ Good    ❏ Satisfactory    ❏ Unsatisfactory    ❏ Poor
Comments: _____
_____

Instructor's Signature _____

*Machining Projects* **Project 2.8** Bored Bar

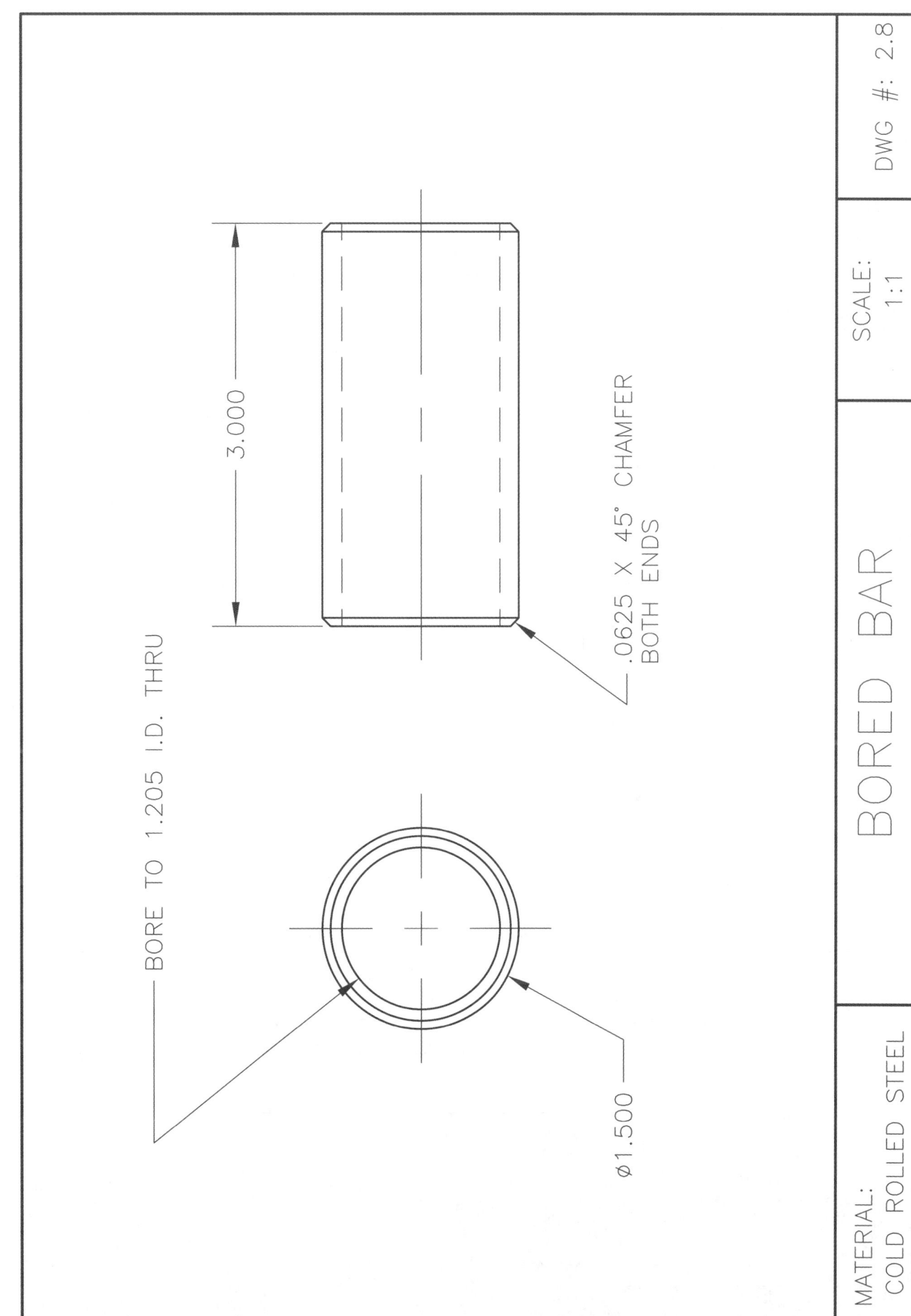

**Project 2.9.1** Tapered Bar  *Machining Projects*

# Project 2.9.1
# Tapered Bar

Name _____  Date _____

Instructor _____  Period _____

## NIMS Duties

*The tasks in this project develop skills related to the following NIMS duties:*

- Duty 1.1  Job Process Planning
- Duty 2.3  Turning Operations: Between Centers Turning
- Duty 2.4  Turning Operations: Chucking
- Duty 2.7a  Grinding Wheel Safety

## Order of Operations

**4.1, 12.1–12.6** ❑ 1. Select, measure, and cut stock plus facing allowance.

**13.11** ❑ 2. Face to specified length.

**14.2** ❑ 3. Using the appropriate taper formulas, calculate the taper per foot (TPF), the taper per inch (TPI), and the angle of taper.

**14.3–14.4** ❑ 4. Using the information calculated in Step 3, determine the correct tailstock set over, adjust the machine, and cut the male taper.

## Notes

_____
_____
_____
_____
_____
_____
_____
_____
_____
_____
_____
_____
_____
_____
_____

---

**Performance Evaluation—Instructor's Use Only**

Project completed on time?  ❑ Yes  ❑ No

If No, which steps were not completed _____

Overall performance on project:

❑ Excellent  ❑ Good  ❑ Satisfactory  ❑ Unsatisfactory  ❑ Poor

Comments: _____
_____

Instructor's Signature _____

Machining Projects  **Project 2.9.1** Tapered Bar

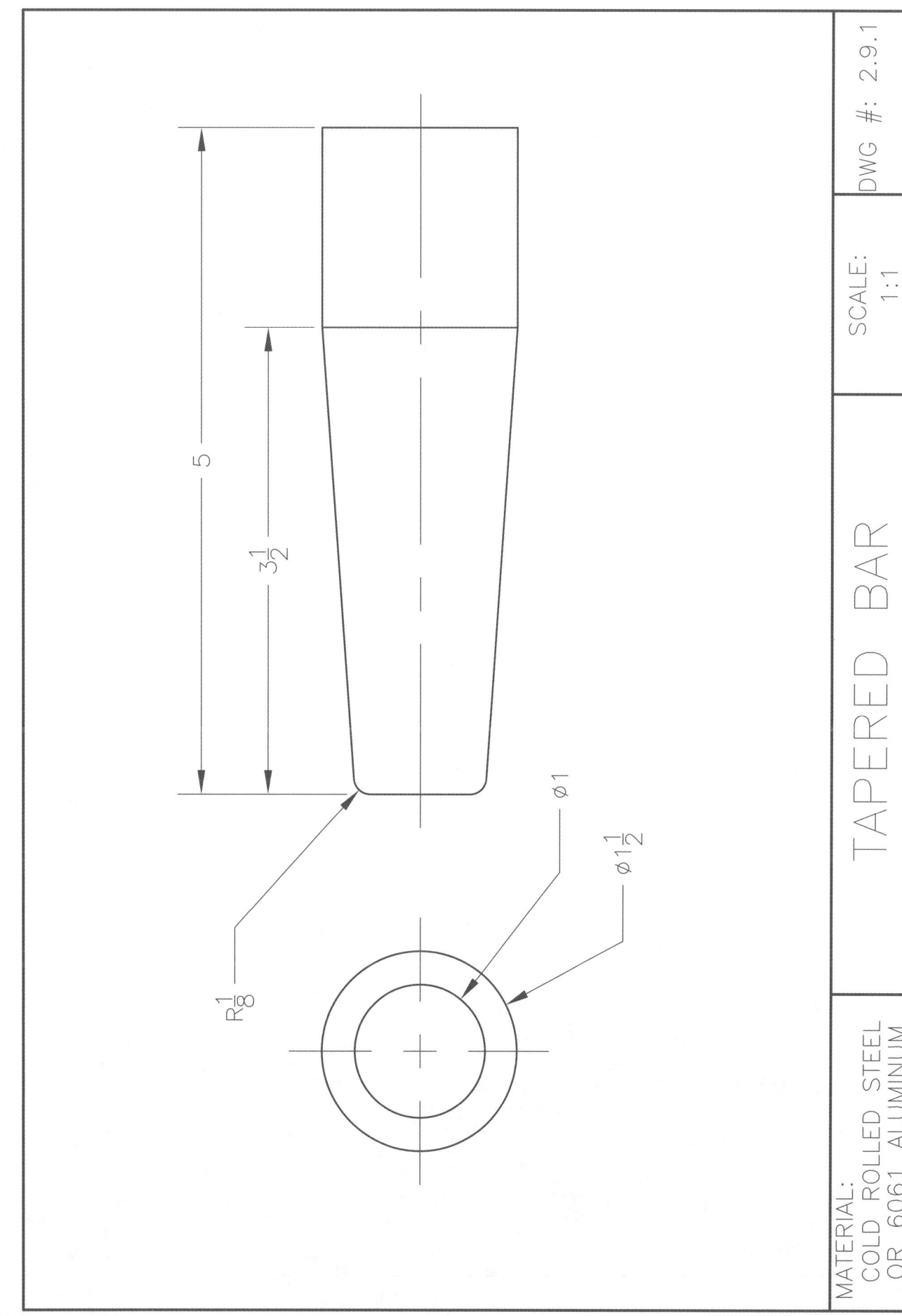

# Project 2.9.2
## Tapered Sleeve

Name _____  Date _____

Instructor _____  Period _____

### NIMS Duties
The tasks in this project develop skills related to the following NIMS duties:
- Duty 1.1 Job Process Planning
- Duty 2.4 Turning Operations: Chucking
- Duty 2.7a Grinding Wheel Safety

## Order of Operations

**4.1, 12.1–12.6** ❏ 1. Select, measure, and cut stock plus facing allowance.

**13.11** ❏ 2. Face to specified length

**14.2** ❏ 3. Using the appropriate taper formulas, calculate the taper per foot (TPF), the taper per inch (TPI), and the angle of taper.

**14.4** ❏ 4. Using either the TPI or the angle of taper, set up the taper attachment of the lathe and cut the female taper.

**14.4** ❏ 5. Set lathe compound and cut chamfers.

Note   The male and female tapered pieces should fit closely upon completion.

### Notes

_____

---

**Performance Evaluation—Instructor's Use Only**

Project completed on time?   ❏ Yes   ❏ No

If No, which steps were not completed _____

Overall performance on project:

❏ Excellent   ❏ Good   ❏ Satisfactory   ❏ Unsatisfactory   ❏ Poor

Comments: _____

Instructor's Signature _____

*Machining Projects* **Project 2.9.2** Tapered Sleeve

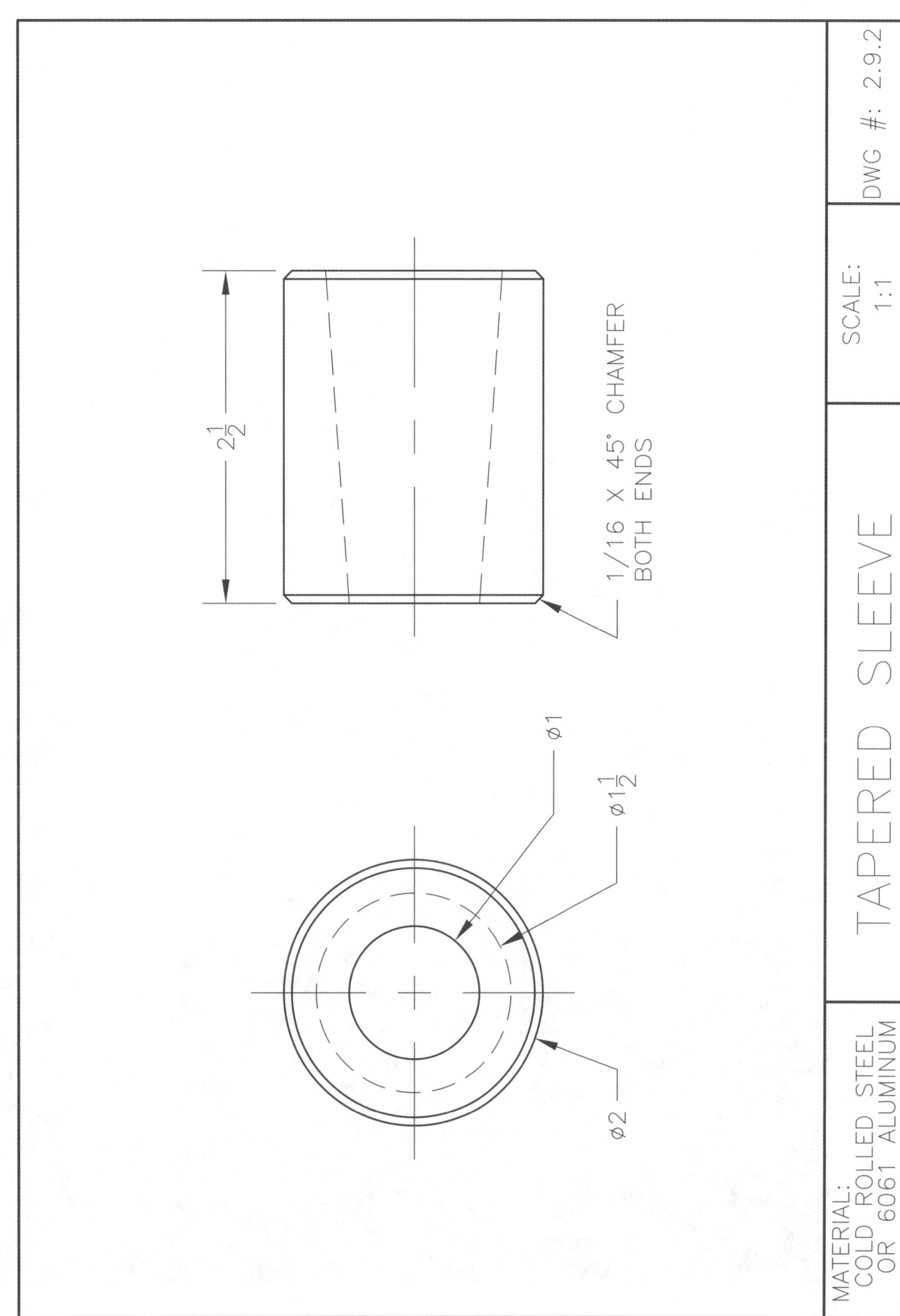

Machining Projects

# Section 3
# Vertical Mill Competency

Machining Projects — Project 3.1 Faced Bar

# Project 3.1
# Faced Bar

Name _____ Date _____

Instructor _____ Period _____

## NIMS Duties

*The tasks in this project develop skills related to the following NIMS duties:*
- Duty 1.1  Job Process Planning
- Duty 2.5  Milling: Square Up a Block
- Duty 2.6  Vertical Milling

## Order of Operations

| | |
|---|---|
| 4.1, 12.1–12.6 ❏ | 1. Select, measure, and cut stock plus facing allowance. |
| 4.5, 18.3 ❏ | 2. Using a dial indicator, trim the head of the vertical mill and indicate the jaws of the mill vise to ensure the squareness of the workpiece. |
| 17.4–17.5 ❏ | 3. Choose and mount an appropriate end mill. |
| 17.8 ❏ | 4. Calculate the correct cutting speed and feed and set the mill accordingly. |
| 17.10, 18.1–18.3 ❏ | 5. Place workpiece in the mill vise, tighten, and set it using a dead blow hammer. |
| 18.1–18.3 ❏ | 6. Using correct procedure for facing to length, cut the workpiece to specifications holding a plus or minus .005 tolerance. |

Note  All measurements should be made with a micrometer or dial caliper.

## Notes

_____
_____
_____
_____
_____
_____
_____
_____
_____
_____
_____
_____

---

**Performance Evaluation—Instructor's Use Only**

Project completed on time?   ❏ Yes   ❏ No
   If No, which steps were not completed _____
Overall performance on project:
   ❏ Excellent   ❏ Good   ❏ Satisfactory   ❏ Unsatisfactory   ❏ Poor
Comments: _____

Instructor's Signature _____

**Project 3.1**  Faced Bar *Machining Projects*

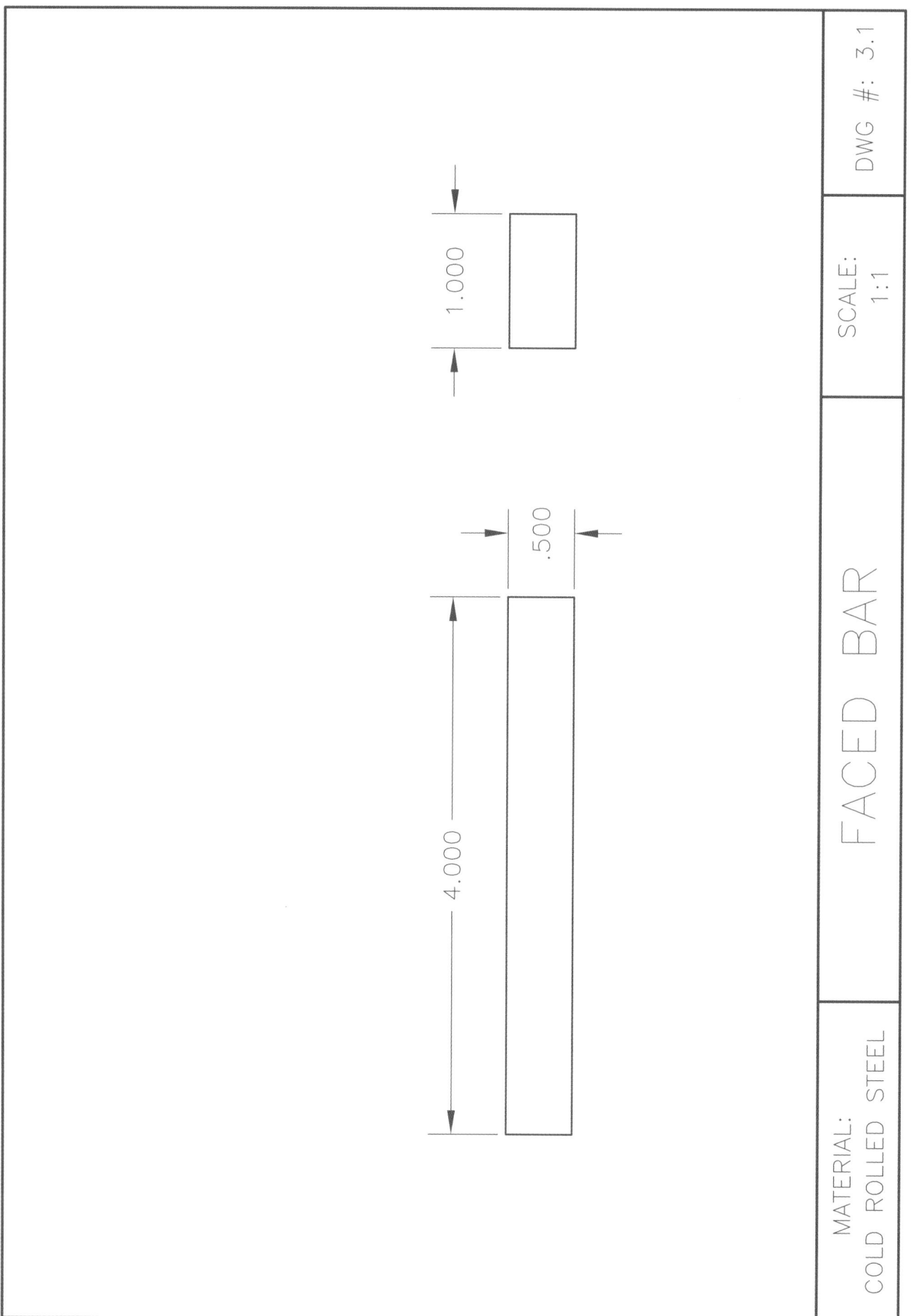

*Machining Projects*          **Project 3.2**     Drilled Plate

# Project 3.2
# Drilled Plate

Name _____ Date _____

Instructor _____ Period _____

## NIMS Duties

*The tasks in this project develop skills related to the following NIMS duties:*
- Duty 1.1    Job Process Planning
- Duty 2.5    Milling: Square Up a Block
- Duty 2.6    Vertical Milling
- Duty 2.8    Drill Press

## Order of Operations

| | | |
|---|---|---|
| 4.1, 12.1–12.6 | ☐ | 1. Select, measure, and cut stock plus facing allowance. |
| 17.8 | ☐ | 2. Calculate the correct cutting speed and feed for the type and size material being used. |
| 18.1–18.3.1 | ☐ | 3. Face the workpiece to the specified dimensions. |
| 5.4 | ☐ | 4. Perform layout work as specified on the print. |
| 10.9 | ☐ | 5. Use a wiggler/center finder to locate the center of each hole before drilling. |
| 10.9 | ☐ | 6. Using the specified drills, drill the holes as indicated on the print. |

**Notes**

_____
_____
_____
_____
_____
_____
_____
_____
_____
_____
_____
_____
_____
_____

---

**Performance Evaluation—Instructor's Use Only**

Project completed on time?    ☐ Yes    ☐ No

     If No, which steps were not completed _____

Overall performance on project:

     ☐ Excellent    ☐ Good    ☐ Satisfactory    ☐ Unsatisfactory    ☐ Poor

Comments: _____

_____

Instructor's Signature _____

**Project 3.2**  Drilled Plate

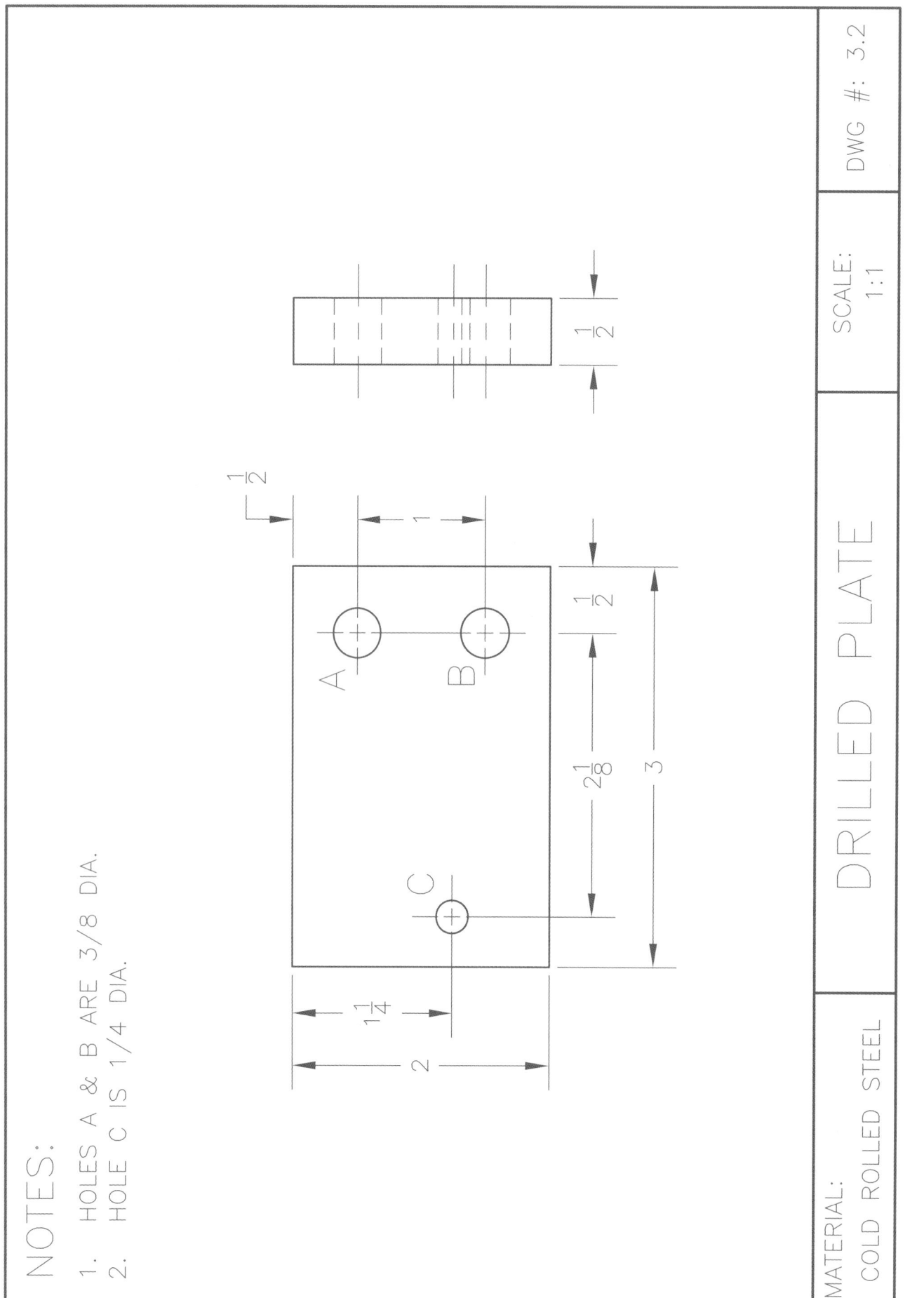

Machining Projects — Project 3.3 — Drilled Plate

# Project 3.3
# Drilled Plate

Name _____ Date _____

Instructor _____ Period _____

## NIMS Duties
*The tasks in this project develop skills related to the following NIMS duties:*
- Duty 1.1    Job Process Planning
- Duty 2.5    Milling: Square Up a Block
- Duty 2.6    Vertical Milling
- Duty 2.8    Drill Press

## Order of Operations

| | | |
|---|---|---|
| 4.1, 12.1–12.6 | ❏ | 1. Select, measure, and cut stock plus facing allowance. |
| 17.8 | ❏ | 2. Calculate the correct speed and feed for the size and type of material and end mill being used. |
| 18.1–18.3 | ❏ | 3. Face the workpiece to the correct dimensions. |
| 18.3 | ❏ | 4. With an edgefinder, accurately locate one end and one edge of the workpiece. |
| 10.6–10.7 | ❏ | 5. Calculate the correct speeds for the specified drill bits and drill the holes indicated on the print. |

Note    All positioning of the work should be done by stepping off the dimensions using the micrometer collars of the mill table.

## Notes

_____
_____
_____
_____
_____
_____
_____
_____
_____
_____
_____
_____

---

**Performance Evaluation—Instructor's Use Only**

Project completed on time? ❏ Yes ❏ No

    If No, which steps were not completed _____

Overall performance on project:
    ❏ Excellent ❏ Good ❏ Satisfactory ❏ Unsatisfactory ❏ Poor

Comments: _____
_____

Instructor's Signature _____

**Project 3.3**  Drilled Plate  *Machining Projects*

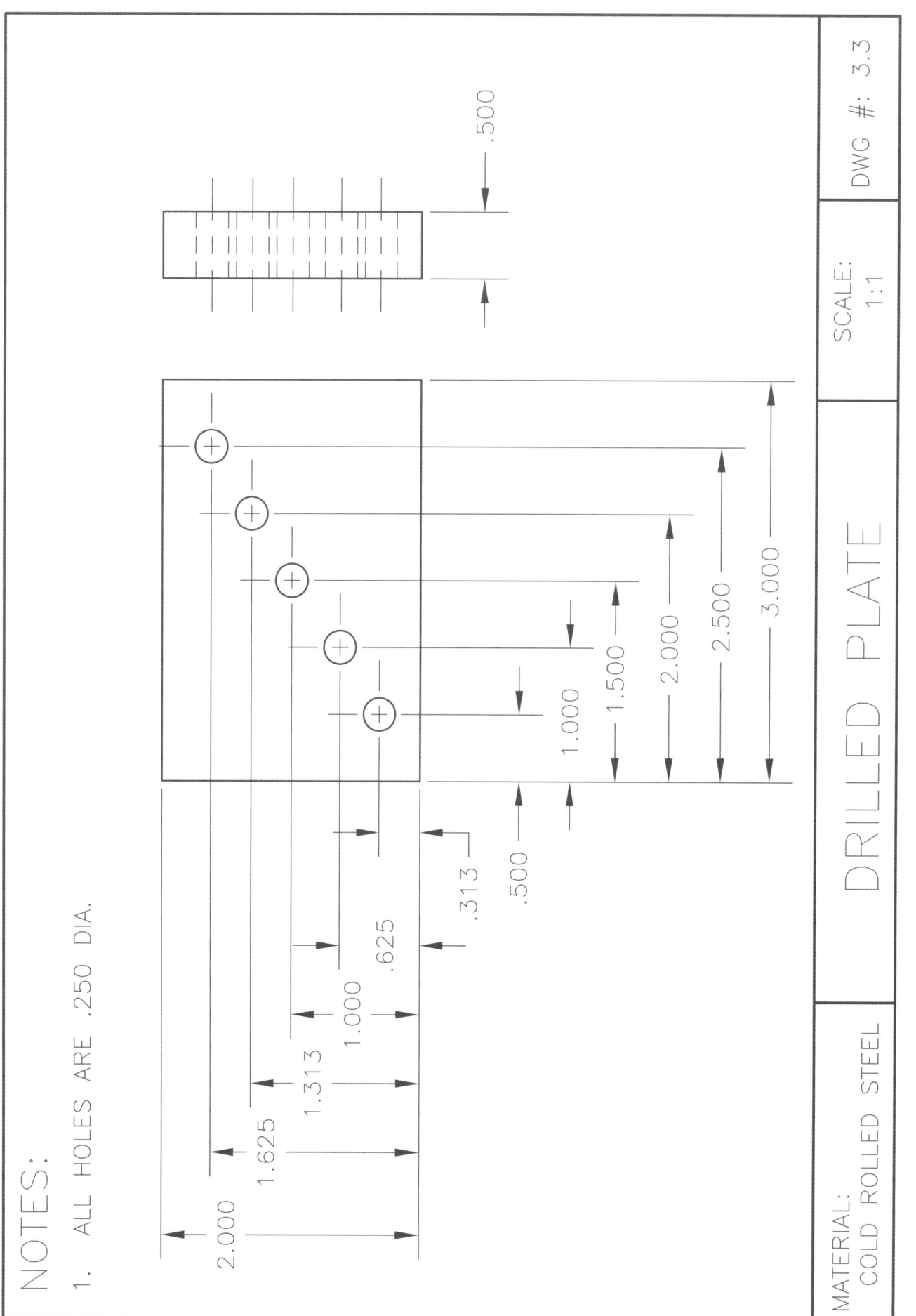

*Machining Projects*      **Project 3.4**     Step Plate

# Project 3.4
# Step Plate

Name _____ Date _____

Instructor _____ Period _____

## NIMS Duties

*The tasks in this project develop skills related to the following NIMS duties:*
- Duty 1.1    Job Process Planning
- Duty 2.5    Milling: Square Up a Block
- Duty 2.6    Vertical Milling

## Order of Operations

| | | |
|---|---|---|
| 4.1, 12.1–12.6 | ❏ | 1. Select, measure, and cut stock plus facing allowance. |
| 17.8 | ❏ | 2. Calculate the correct speed and feed for the type and size material and end mill being used. |
| 18.1–18.3 | ❏ | 3. Machine the workpiece to correct overall dimensions. |
| 18.3 | ❏ | 4. Locate the first end by touching off with the end mill. |
| 18.3 | ❏ | 5. Cut the step to the specified depth and width. |
| 18.3 | ❏ | 6. Locate the opposite end of the workpiece with an edge finder. |
| 18.3 | ❏ | 7. Cut the step to the specified dimensions. |

## Notes

_____
_____
_____
_____
_____
_____
_____
_____
_____
_____
_____
_____
_____

**Performance Evaluation—Instructor's Use Only**
Project completed on time?    ❏ Yes    ❏ No
    If No, which steps were not completed _____
Overall performance on project:
    ❏ Excellent    ❏ Good    ❏ Satisfactory    ❏ Unsatisfactory    ❏ Poor
Comments: _____
_____

Instructor's Signature _____

Copyright by Goodheart-Willcox Co., Inc.

**Project 3.4** Step Plate *Machining Projects*

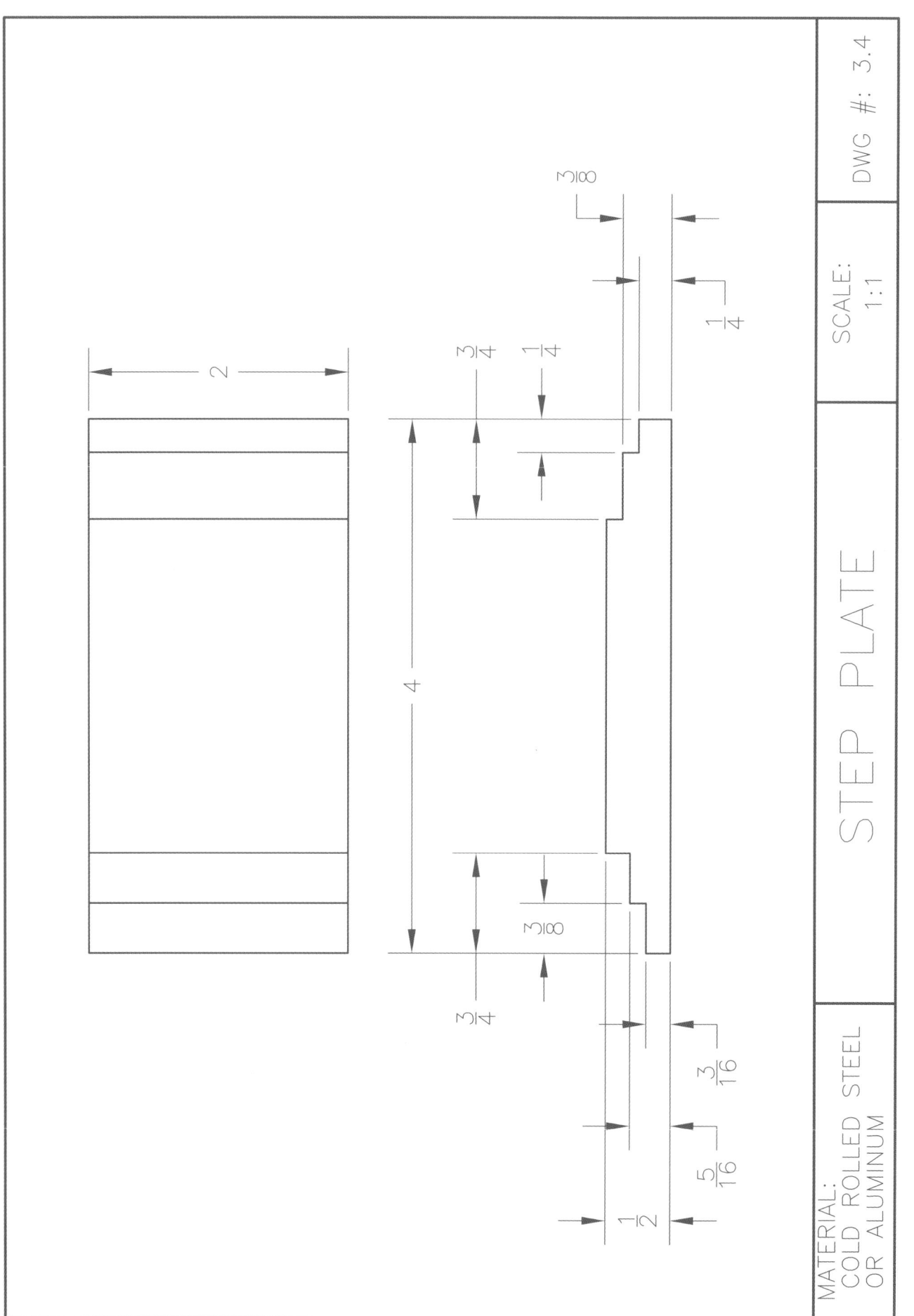

Machining Projects | Project 3.5 Slotted Block or Bar

# Project 3.5
# Slotted Block or Bar

Name _____ Date _____

Instructor _____ Period _____

## NIMS Duties

*The tasks in this project develop skills related to the following NIMS duties:*
- Duty 1.1   Job Process Planning
- Duty 2.5   Milling: Square Up a Block
- Duty 2.6   Vertical Milling

## Order of Operations

| | | |
|---|---|---|
| 4.1, 12.1–12.6 | ❏ | 1. Select, measure, and cut stock plus facing allowance. |
| 17.8 | ❏ | 2. Calculate the correct speed and feed for the type and size material and end mill to be used. |
| 18.1–18.3 | ❏ | 3. Face the workpiece to the specified dimensions. |
| 18.3 | ❏ | 4. Locate edges using an edge finder. |
| 18.3 | ❏ | 5. Locate the top surface by touching off with an end mill. |
| 18.3 | ❏ | 6. Position the end mill for the beginning of the slot. |
| 18.3 | ❏ | 7. Using a combination of the quill stop and the vertical micrometer table adjustment, set the correct depth of cut. |
| 18.3 | ❏ | 8. Down-feed the end mill (it must be a two-flute) to the correct depth and complete the cut to specifications. |

## Notes

_____
_____
_____
_____
_____
_____
_____
_____

---

**Performance Evaluation—Instructor's Use Only**

Project completed on time?   ❏ Yes   ❏ No
    If No, which steps were not completed _____
Overall performance on project:
    ❏ Excellent   ❏ Good   ❏ Satisfactory   ❏ Unsatisfactory   ❏ Poor
Comments: _____
_____

Instructor's Signature _____

**Project 3.5**  Slotted Block or Bar  *Machining Projects*

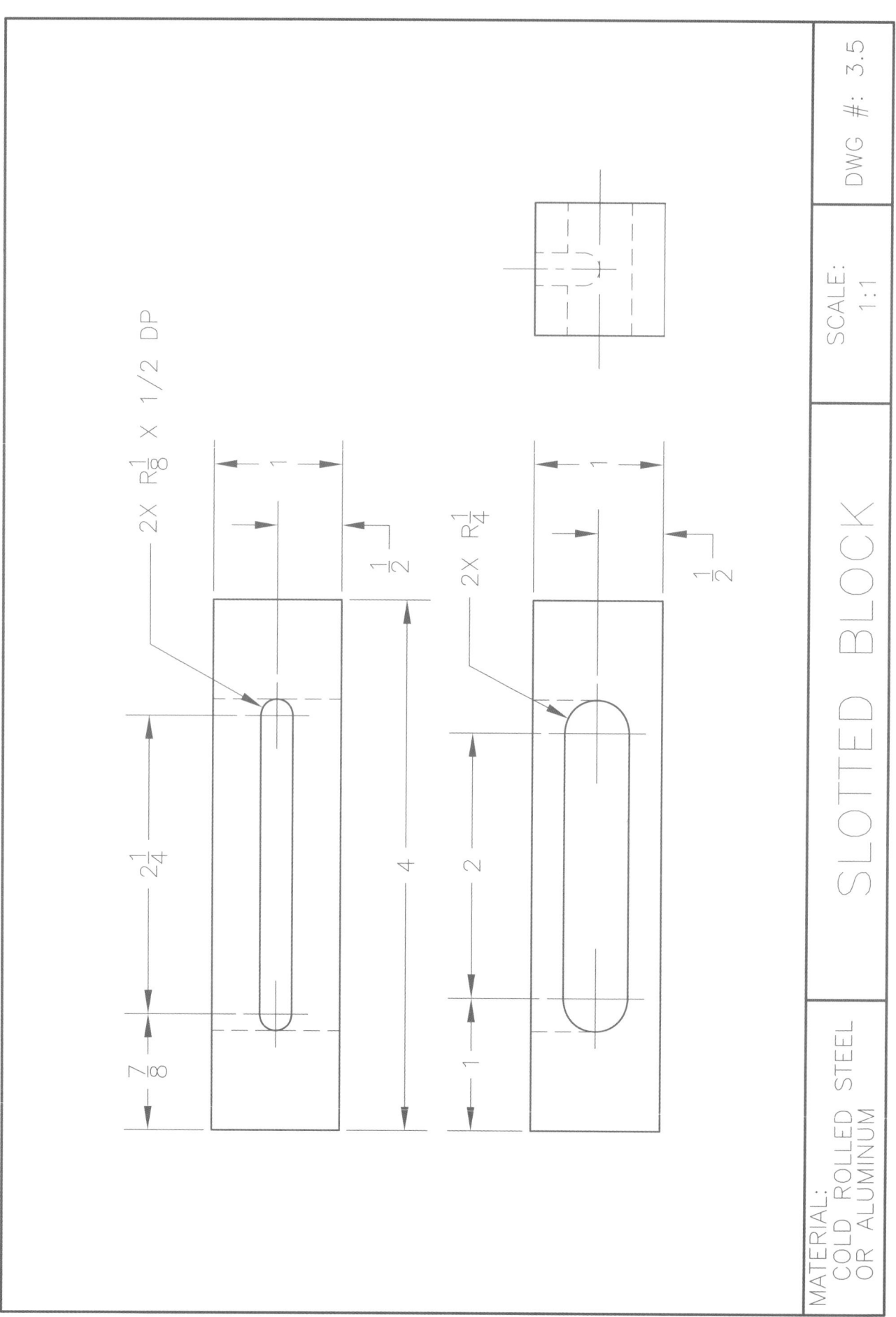

# Project 3.6
## Multi-Operations Bar

Name _____  Date _____

Instructor _____  Period _____

### NIMS Duties
*The tasks in this project develop skills related to the following NIMS duties:*
- ○ Duty 1.1  Job Process Planning
- ○ Duty 2.5  Milling: Square Up a Block
- ○ Duty 2.6  Vertical Milling
- ○ Duty 2.8  Drill Press

## Order of Operations

| | | |
|---|---|---|
| **4.1, 12.1–12.6** | ❏ | 1. Select, measure, and cut stock plus facing allowance. |
| **17.8** | ❏ | 2. Calculate appropriate speeds and feeds for each of the cutting tools being used. |
| **18.1–18.3** | ❏ | 3. Machine to specified length and width. |
| **17.5** | ❏ | 4. Fly cut to specified thickness. |
| **10.9–10.13, 18.3** | ❏ | 5. Use an edge finder to locate the centers of holes, then drill, countersink to the major diameter of the thread, and power tap. |

## Notes

_____
_____
_____
_____
_____
_____
_____
_____
_____
_____
_____
_____
_____
_____

---

**Performance Evaluation—Instructor's Use Only**

Project completed on time?  ❏ Yes  ❏ No

  If No, which steps were not completed _____

Overall performance on project:

  ❏ Excellent  ❏ Good  ❏ Satisfactory  ❏ Unsatisfactory  ❏ Poor

Comments: _____
_____

Instructor's Signature _____

**Project 3.6** Multi-Operations Bar

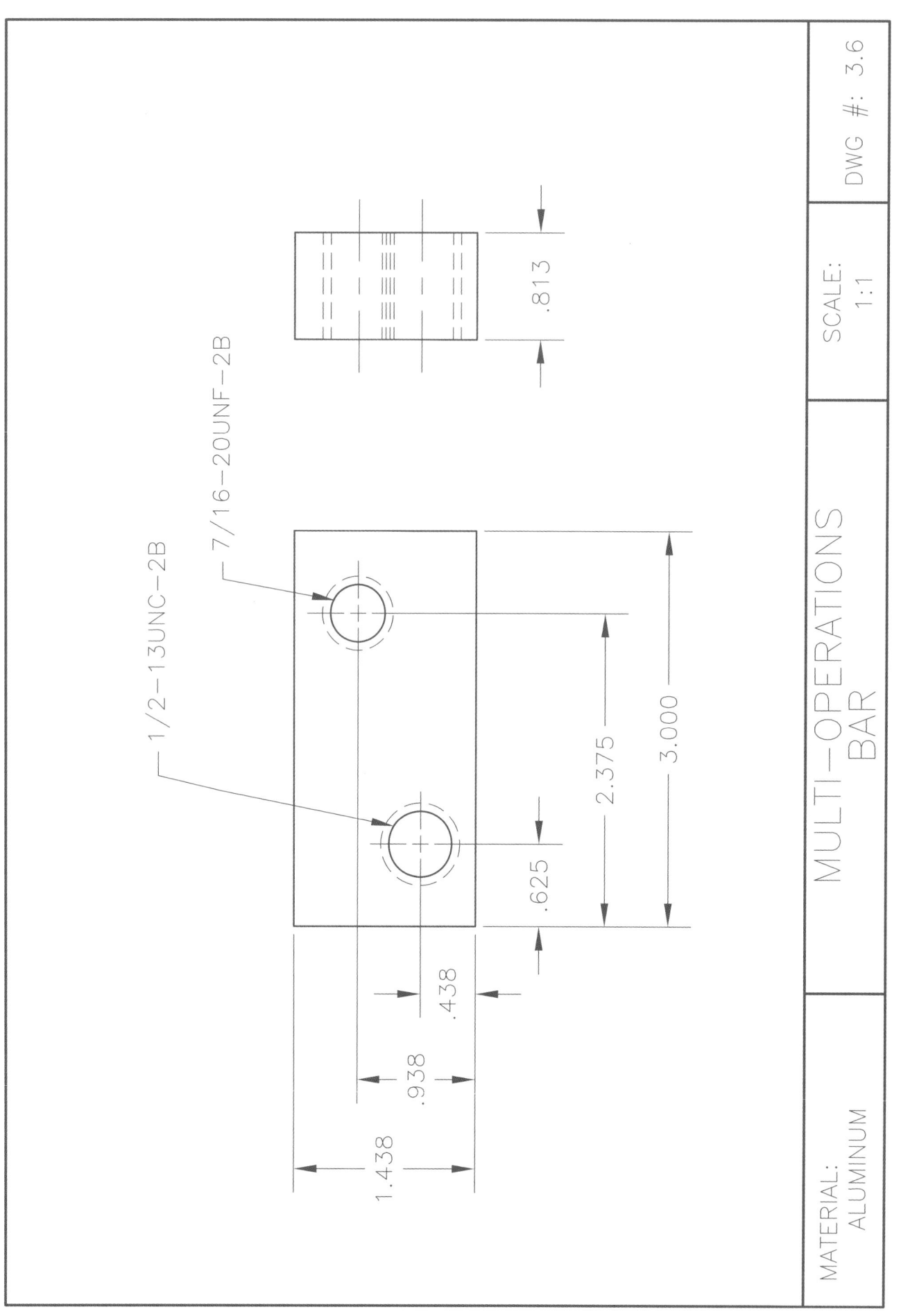

# Project 3.7
## Angle Bar

Name _____  Date _____

Instructor _____  Period _____

### NIMS Duties

*The tasks in this project develop skills related to the following NIMS duties:*
- Duty 1.1    Job Process Planning
- Duty 2.5    Milling: Square Up a Block
- Duty 2.6    Vertical Milling

## Order of Operations

| | | |
|---|---|---|
| **4.1, 12.1–12.6** | ❑ | 1. Select, measure, and cut stock plus facing allowance. |
| **17.8** | ❑ | 2. Calculate the correct cutting speeds and feeds for each of the cutting tools to be used. |
| **18.1–18.3** | ❑ | 3. Face the workpiece to the correct overall length. |
| **18.3** | ❑ | 4. Adjust the head of the vertical mill to the specified angle and cut as indicated on the drawing. |
| **18.3** | ❑ | 5. Using a dial indicator, trim the head of the mill to square. |
| **18.3** | ❑ | 6. Adjust the mill vise to the specified angle and make the cut as indicated on the drawing. |

### Notes

_____
_____
_____
_____
_____
_____
_____
_____
_____
_____

---

**Performance Evaluation—Instructor's Use Only**

Project completed on time?    ❑ Yes    ❑ No

If No, which steps were not completed _____

Overall performance on project:
     ❑ Excellent    ❑ Good    ❑ Satisfactory    ❑ Unsatisfactory    ❑ Poor

Comments: _____
_____

Instructor's Signature _____

**Project 3.7** Angle Bar

*Machining Projects*

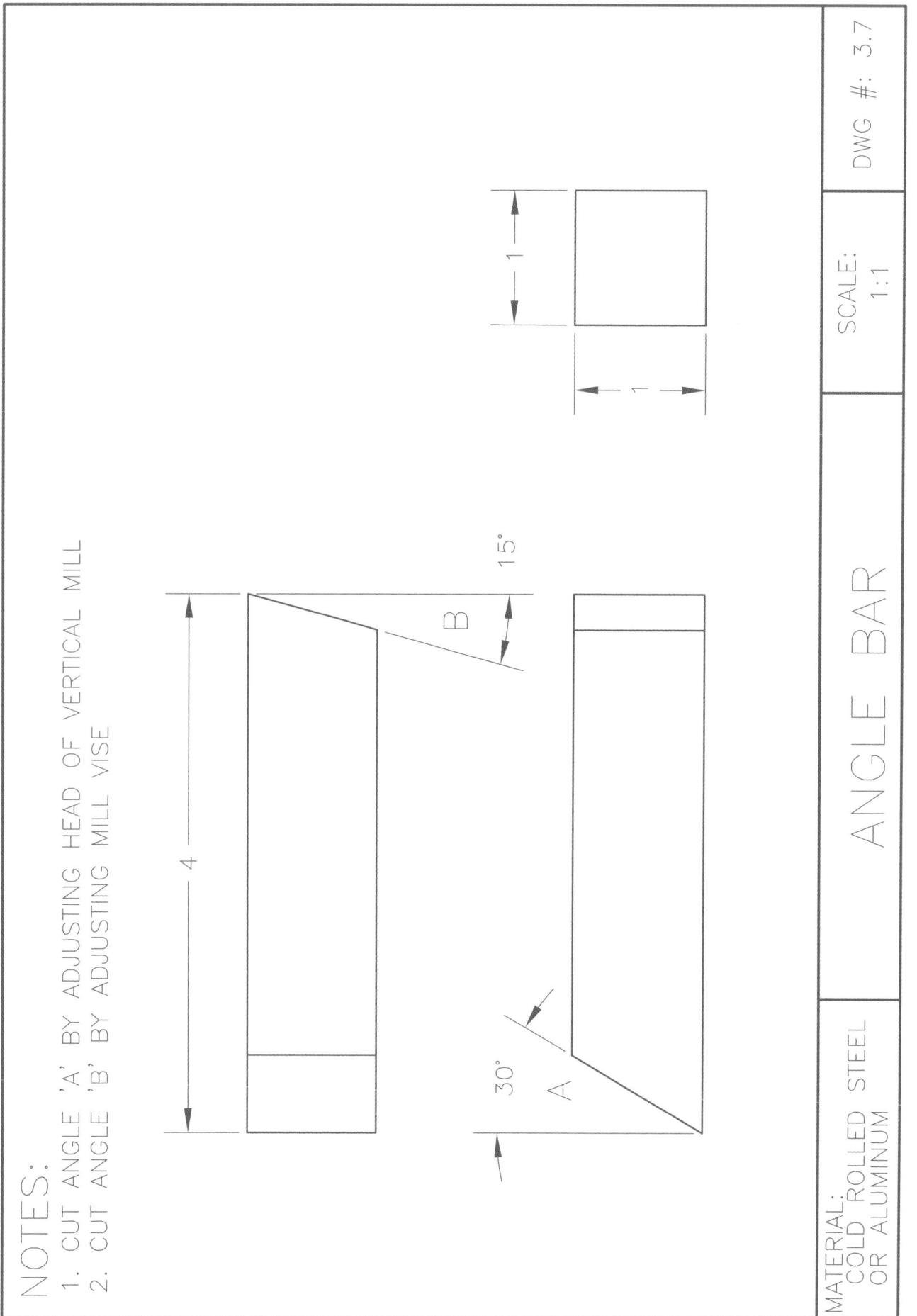

Machining Projects  Project 3.8  Round Bar

# Project 3.8
# Round Bar

Name _____  Date _____

Instructor _____  Period _____

**NIMS Duties**

*The tasks in this project develop skills related to the following NIMS duties:*
- Duty 1.1  Job Process Planning
- Duty 2.5  Milling: Square Up a Block
- Duty 2.6  Vertical Milling

## Order of Operations

| | | |
|---|---|---|
| 4.1 | ❑ | 1. Measure and cut stock plus facing allowance. |
| 13.11, 18.1–18.3 | ❑ | 2. Face to the specified length (lathe or mill). |
| 17.10, 18.3 | ❑ | 3. Mount workpiece solidly in the mill vise. |
| 18.3 | ❑ | 4. With an edge finder, touch off on a side, then center the spindle over the center of the piece. |
| 17.4–17.5 | ❑ | 5. Choose and mount an appropriate end mill. |
| 18.3 | ❑ | 6. Raise the mill table until the end mill touches off on the top surface of the workpiece. |
| 18.3 | ❑ | 7. Zero the vertical adjustment collar on the mill and adjust to the correct depth of cut. |
| 18.3 | ❑ | 8. Cut the flat section on the bar. |
| 18.3 | ❑ | 9. Locate the end edge with an edge finder. |
| 18.3 | ❑ | 10. Center the spindle over the edge, zero the micrometer collar, and then install a 1/4″ end mill. |
| 18.3 | ❑ | 11. Raise the mill table until the end mill touches off on the top surface of the workpiece. |
| 18.3 | ❑ | 12. Zero the vertical adjustment on the mill table and raise the table to the specified depth of cut. |
| 18.3 | ❑ | 13. Cut the keyway to the specified length and depth. |

**Notes**

_____
_____
_____
_____

---

**Performance Evaluation—Instructor's Use Only**
Project completed on time?   ❑ Yes   ❑ No
    If No, which steps were not completed _____
Overall performance on project:
    ❑ Excellent   ❑ Good   ❑ Satisfactory   ❑ Unsatisfactory   ❑ Poor
Comments: _____
_____

Instructor's Signature _____

Copyright by Goodheart-Willcox Co., Inc.

**Project 3.8** Round Bar

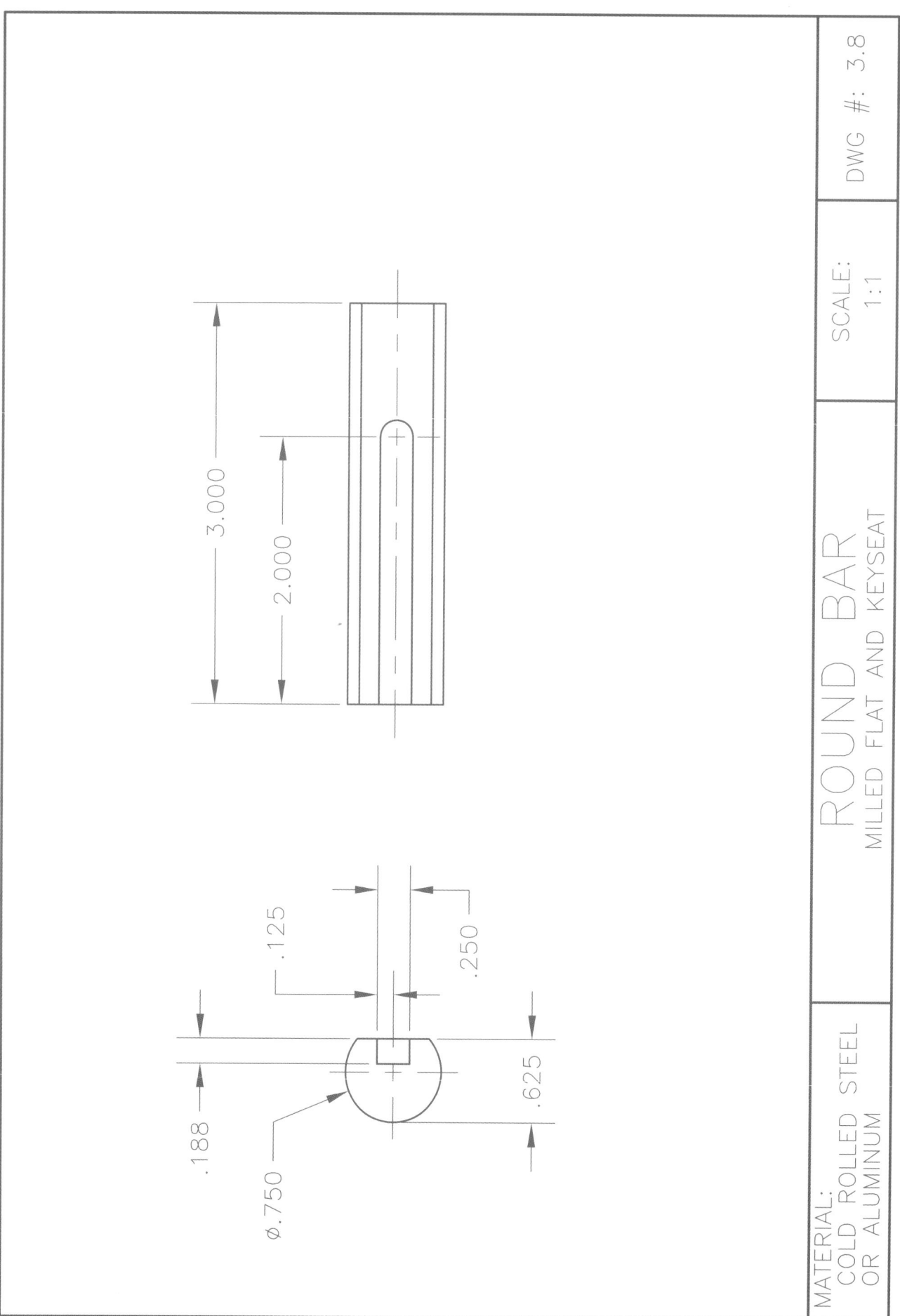

# Project 3.9
# Mill Block—Drilled and Bored Holes

Name _____ Date _____

Instructor _____ Period _____

## NIMS Duties
*The tasks in this project develop skills related to the following NIMS duties:*
- Duty 1.1  Job Process Planning
- Duty 2.5  Milling: Square Up a Block
- Duty 2.6  Vertical Milling
- Duty 2.8  Drill Press

## Order of Operations

| | | |
|---|---|---|
| **4.1, 12.1–12.6** | ❏ | 1. Measure and cut stock plus facing allowance. |
| **4.5, 18.3** | ❏ | 2. Check the alignment of the vertical mill head and vise with a dial indicator. |
| **17.4, 17.5** | ❏ | 3. Choose and mount an appropriate end mill. |
| **17.8** | ❏ | 4. Calculate the correct cutting speed and feed for the type and size end mill to be used. |
| **18.3** | ❏ | 5. Using the correct cutting sequence, mill the workpiece to the dimensions specified on the drawing. |
| **18.3** | ❏ | 6. Use an edge finder to accurately locate the center of the longitudinal hole. |
| **10.9** | ❏ | 7. Drill the 1″ diameter through hole. |
| **18.3** | ❏ | 8. Set up the boring head in the vertical mill and bore the hole to the specified depth and diameter. |
| **10.9, 18.3** | ❏ | 9. Reposition the workpiece in the mill vise, find the center for the transverse hole, and drill 1″. |
| **18.3** | ❏ | 10. Set up the boring head in the vertical mill and bore the hole to the specified depth and diameter. |
| | ❏ | 11. Remove burrs and finish the piece to your instructor's specifications. |

## Notes

_____
_____
_____
_____
_____

---

**Performance Evaluation—Instructor's Use Only**
Project completed on time?  ❏ Yes  ❏ No
  If No, which steps were not completed _____
Overall performance on project:
  ❏ Excellent  ❏ Good  ❏ Satisfactory  ❏ Unsatisfactory  ❏ Poor
Comments: _____
_____

Instructor's Signature _____

**Project 3.9**  Mill Block—Drilled and Bored Holes  *Machining Projects*

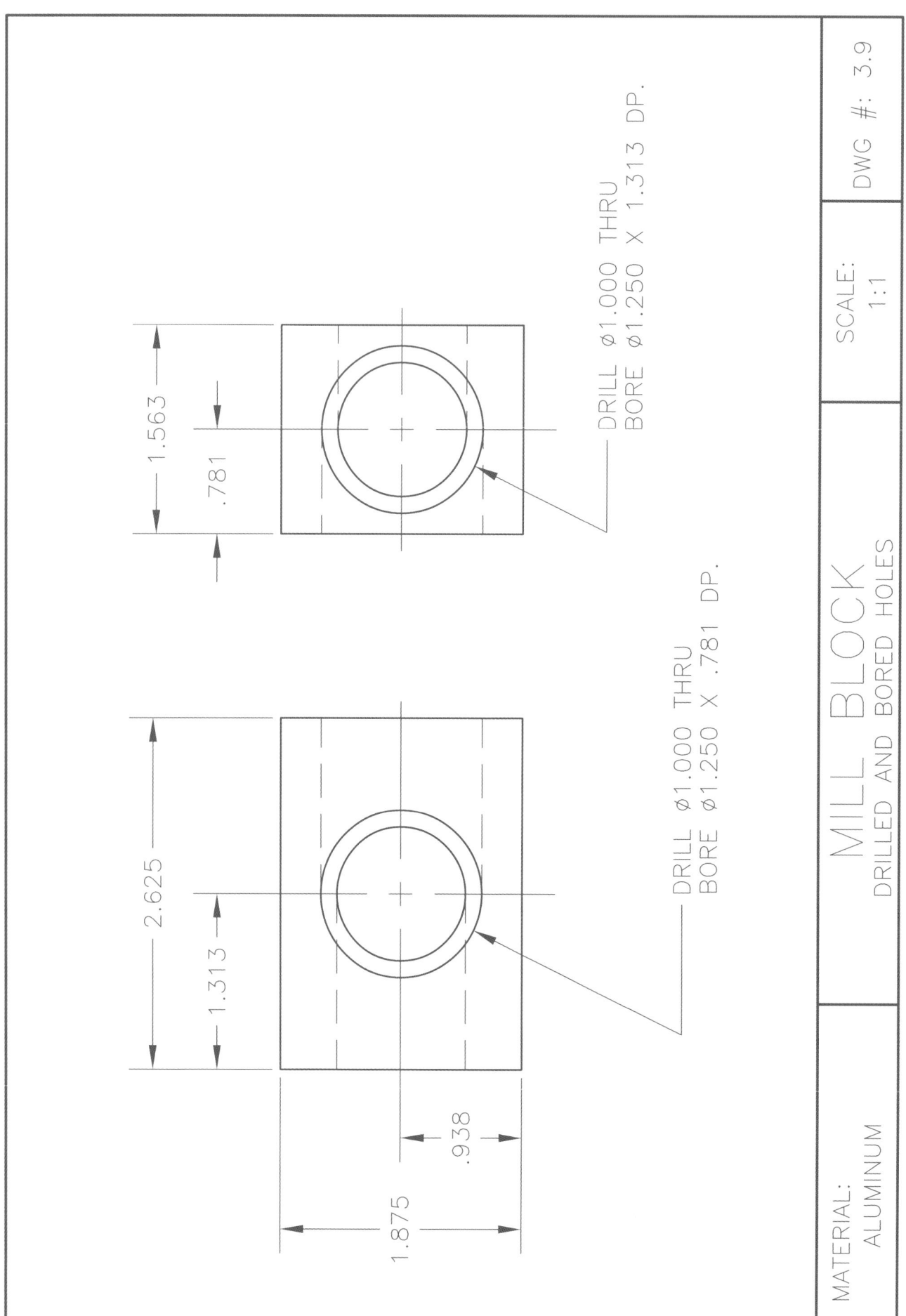

*Machining Projects*      **Project 3.10**    Fluted Round Bar

# Project 3.10
# Fluted Round Bar

Name _____ Date _____

Instructor _____ Period _____

## NIMS Duties

*The tasks in this project develop skills related to the following NIMS duties:*
- Duty 1.1   Job Process Planning
- Duty 2.4   Turning Operations: Chucking
- Duty 2.5   Milling: Square Up a Block
- Duty 2.6   Vertical Milling
- Duty 2.8   Drill Press

## Order of Operations

| Ref | Step |
|---|---|
| 4.1, 12.1–12.6 | 1. Measure and cut stock plus facing allowance. |
| 13.11 | 2. Face the bar to print specifications and center drill (lathe). |
| 17.10 | 3. Install and indicate dividing head on the vertical mill. |
|  | 4. Secure the workpiece in the dividing head. See reference above. |
| 18.3 | 5. Using an edge finder, locate the center and end of the workpiece. |
| 17.8 | 6. Install appropriate end mill and calculate correct cutting speed and feed for the size and type of cutter being used. |
| 18.3 | 7. Using the vertical adjustment of the mill, touch off, set zero, and adjust to the specified depth of cut. |
| 18.3 | 8. Make the first cut, then rotate the workpiece to the next position. |
|  | 9. Repeat Step 8 until all flutes are cut. |

## Notes

_____
_____
_____
_____
_____
_____
_____
_____
_____

---

**Performance Evaluation—Instructor's Use Only**

Project completed on time? ❏ Yes ❏ No

If No, which steps were not completed _____

Overall performance on project:
❏ Excellent   ❏ Good   ❏ Satisfactory   ❏ Unsatisfactory   ❏ Poor

Comments: _____

Instructor's Signature _____

# Project 3.10 Fluted Round Bar

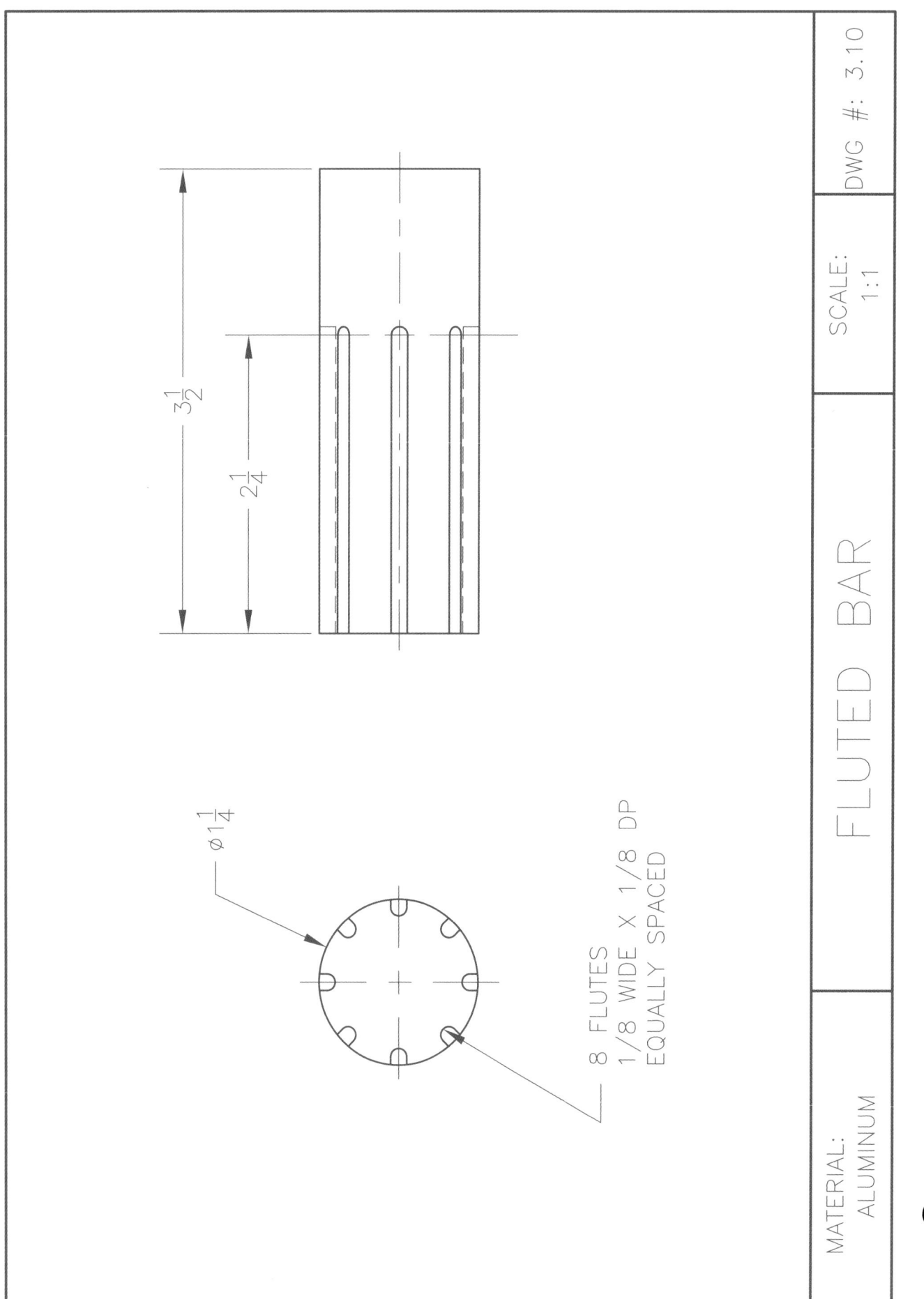

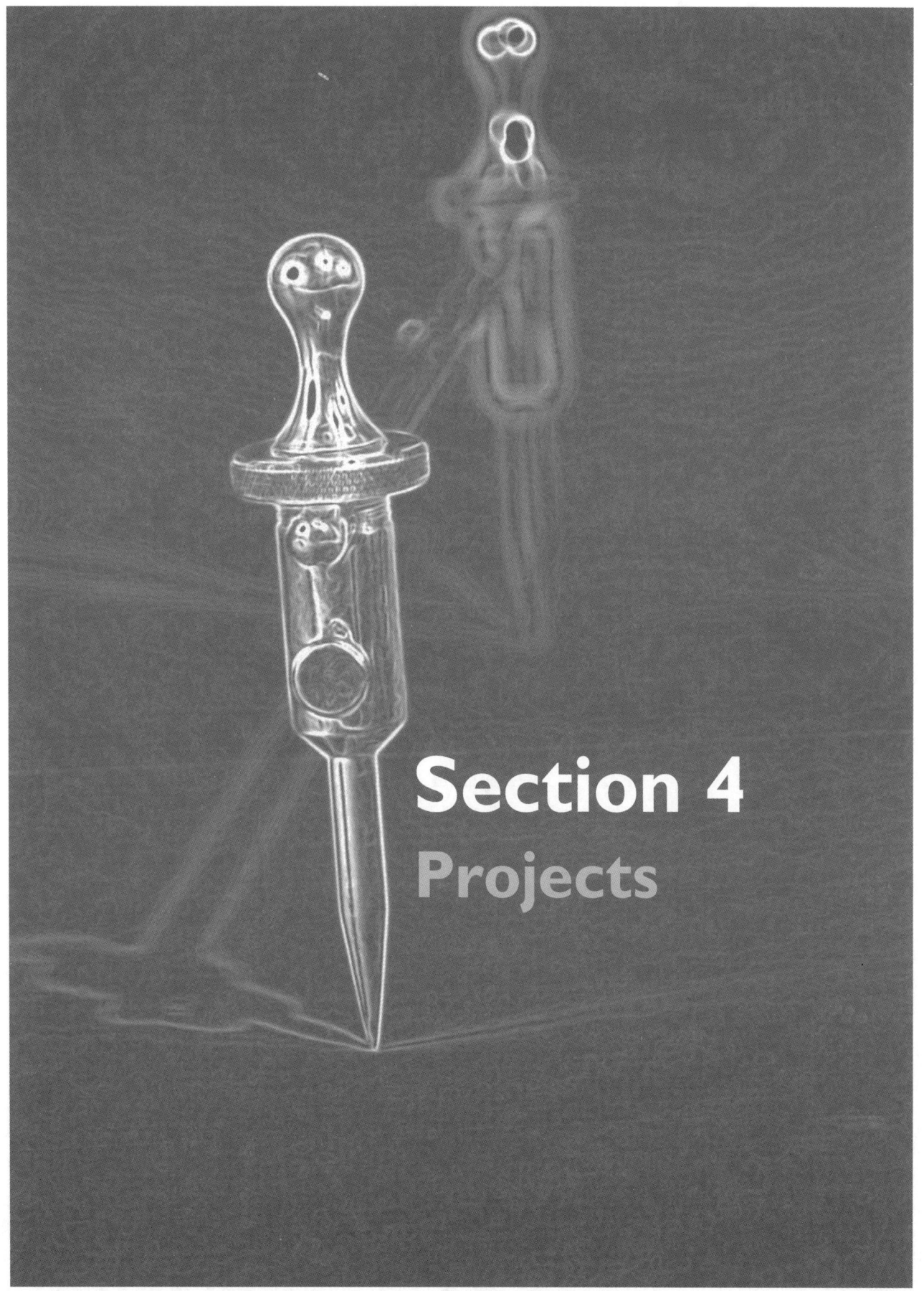

# Section 4
## Projects

# Project 4.1
## Barstock Spindle Stop Adapter for Lathe

Name _____  Date _____

Instructor _____  Period _____

## Order of Operations

- ❏ 1. Measure and cut stock plus facing allowance.
- ❏ 2. Face to the specified length.
- ❏ 3. Center drill, drill .875" through.
- ❏ 4. Set up boring bar and bore through to a slip fit for 1.000" diameter barstock (about 1.020).
- ❏ 5. Using the boring bar again, bore one end of the piece ⌀1.875" × 1.375" deep.

    Note   The exact diameter of the outside end of the lathe spindle should be determined before performing this operation.

- ❏ 6. Turn the workpiece around and turn down the end to the length and diameter specified on the print.
- ❏ 7. Set up a dividing head on the vertical mill and drill three equally spaced holes in the locations indicated, then tap 3/8-24UNF.
- ❏ 8. Adjust the compound rest of the lathe and cut the 45° angle.
- ❏ 9. Break all edges and deburr all holes.

## Notes

_____
_____
_____
_____
_____
_____
_____
_____
_____
_____

---

**Performance Evaluation—Instructor's Use Only**

Project completed on time?   ❏ Yes   ❏ No
    If No, which steps were not completed _____
Overall performance on project:
    ❏ Excellent   ❏ Good   ❏ Satisfactory   ❏ Unsatisfactory   ❏ Poor
Comments: _____
_____

Instructor's Signature _____

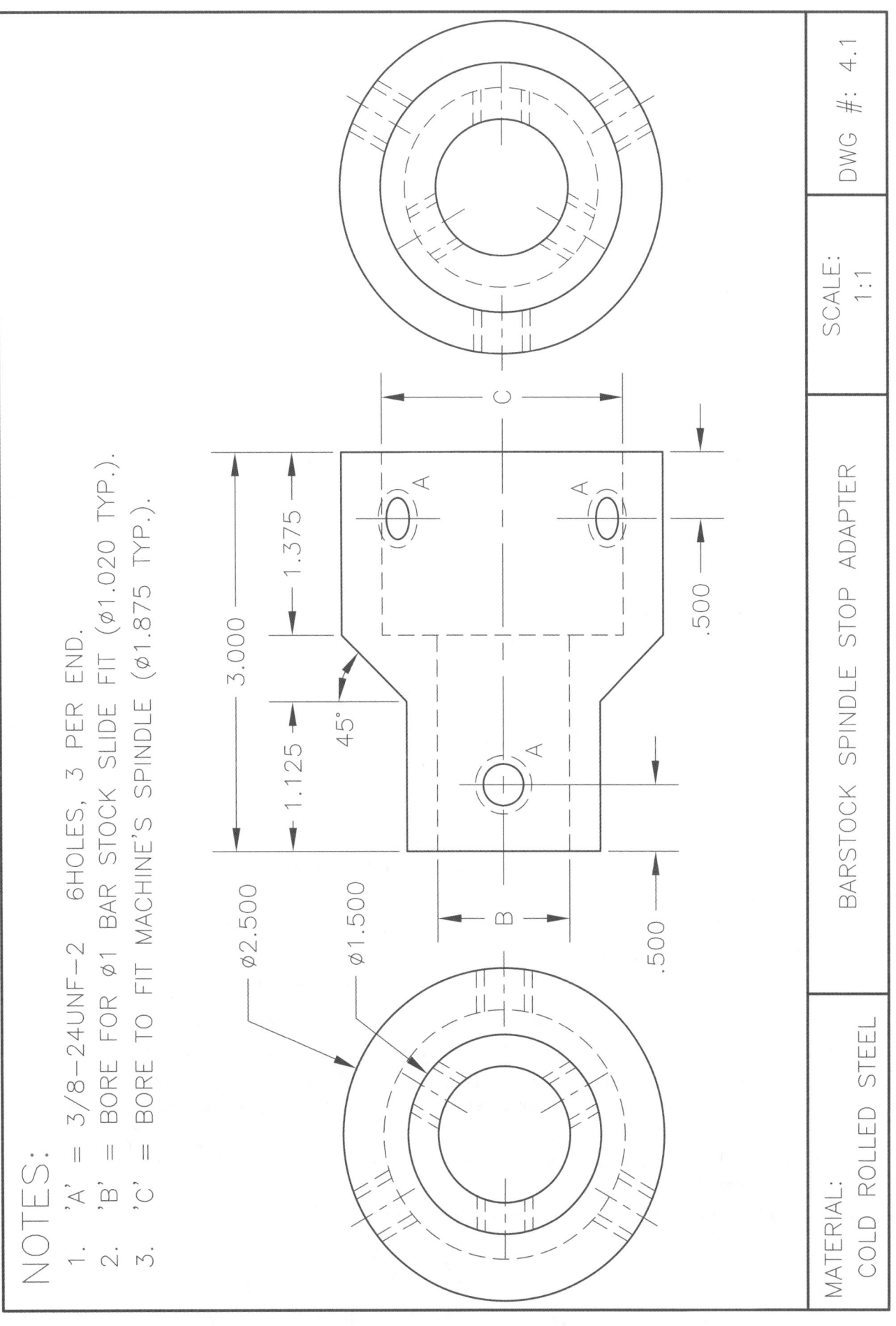

# Project 4.2
## Ball Peen Hammer

Name _____  Date _____

Instructor _____  Period _____

## Order of Operations

### Handle and End Cap

☐ 1. Measure and cut stock. Handle and cap length plus facing allowance for each should be added together when cutting stock (approximately 12″).

☐ 2. Face both ends of the piece just enough to make them straight, smooth, and square.

☐ 3. Center drill both ends.

☐ 4. Drill one end 1/2″ × 5/8″ deep.

☐ 5. Knurl approximately 6″ of this end of the workpiece. (coarse knurl)

☐ 6. Turn the outside of this end to 3/4″, cut or file a chamfer on the edge, and thread 3/4-16.

☐ 7. Cut a 1/16″ wide × 1/16″ deep groove at the shoulder to terminate the threads cleanly.

☐ 8. Cut or part off 1 1/8″ of this end to make the cap.

☐ 9. Turn the cap piece around and face to the specified length. A piece of scrap stock that has been faced and drilled and tapped 3/4-16 can be chucked in the lathe and the cap screwed into it for facing to avoid damaging the knurl.

☐ 10. Turn the end down just enough to remove the knurl and cut or file the chamfer.

☐ 11. Face the handle piece to the length specified on the print and center drill.

☐ 12. Drill the knurled end 11/16″ diameter × 4 5/8″ deep. The use of a collet chuck is recommended for this operation to avoid damaging the knurl. If a collet chuck is not available, a split sleeve with a hole just large enough to slip over the knurled portion of the handle can be machined.

☐ 13. Tap this end of the handle 3/4-16 × 3/4 deep.

☐ 14. Turn the end to the diameter and length indicated on the print.

☐ 15. Reverse the handle in the lathe, chuck it by the smooth section, and engage a tailstock mounted live center in the end.

☐ 16. Turn this end down to 5/8″ diameter × 1 3/4 long.

☐ 17. Turn the end down to .490″ (to ease the threading process) × 1/2″ long.

☐ 18. With a die and die stock, thread the end of the handle 1/2-13.

☐ 19. Cut a 1/16″ wide × 1/16″ deep groove to terminate the threads cleanly.

☐ 20. Using the *Machining Fundamentals* textbook and *Machinery's Handbook* as references, calculate the tailstock setover, taper per inch, taper per foot, and degree of taper for both of the tapers shown.

☐ 21. Offset the tailstock or, if using a taper attachment, set the correct taper per inch or degree of taper.

☐ 22. Cut the tapers as they are shown on the print.

## Hammer Head

- ❏ 1. Measure and cut stock plus facing allowance. (While steel is normally used for the hammer head, polished brass adds greatly to the appearance of the assembled hammer.)
- ❏ 2. Face hammer head to the correct overall length.
- ❏ 3. Set up the vertical mill, place workpiece in the mill vise and locate the correct center location for the 1/2-13 hole.
- ❏ 4. Drill the hole to the specified depth, counterbore, countersink the inside edge of the hole, and tap 1/2-13.
- ❏ 5. Using a radius-ground lathe tool bit, cut the two grooves as shown.
- ❏ 6. Using an outside radius-ground tool bit, rough cut the ball, then file finish.
- ❏ 7. Sand and polish all parts and assemble.

## Notes

---

**Performance Evaluation—Instructor's Use Only**

Project completed on time?  ❏ Yes  ❏ No
  If No, which steps were not completed _____
Overall performance on project:
  ❏ Excellent  ❏ Good  ❏ Satisfactory  ❏ Unsatisfactory  ❏ Poor
Comments: _____

Instructor's Signature _____

# Project 4.2 Ball Peen Hammer

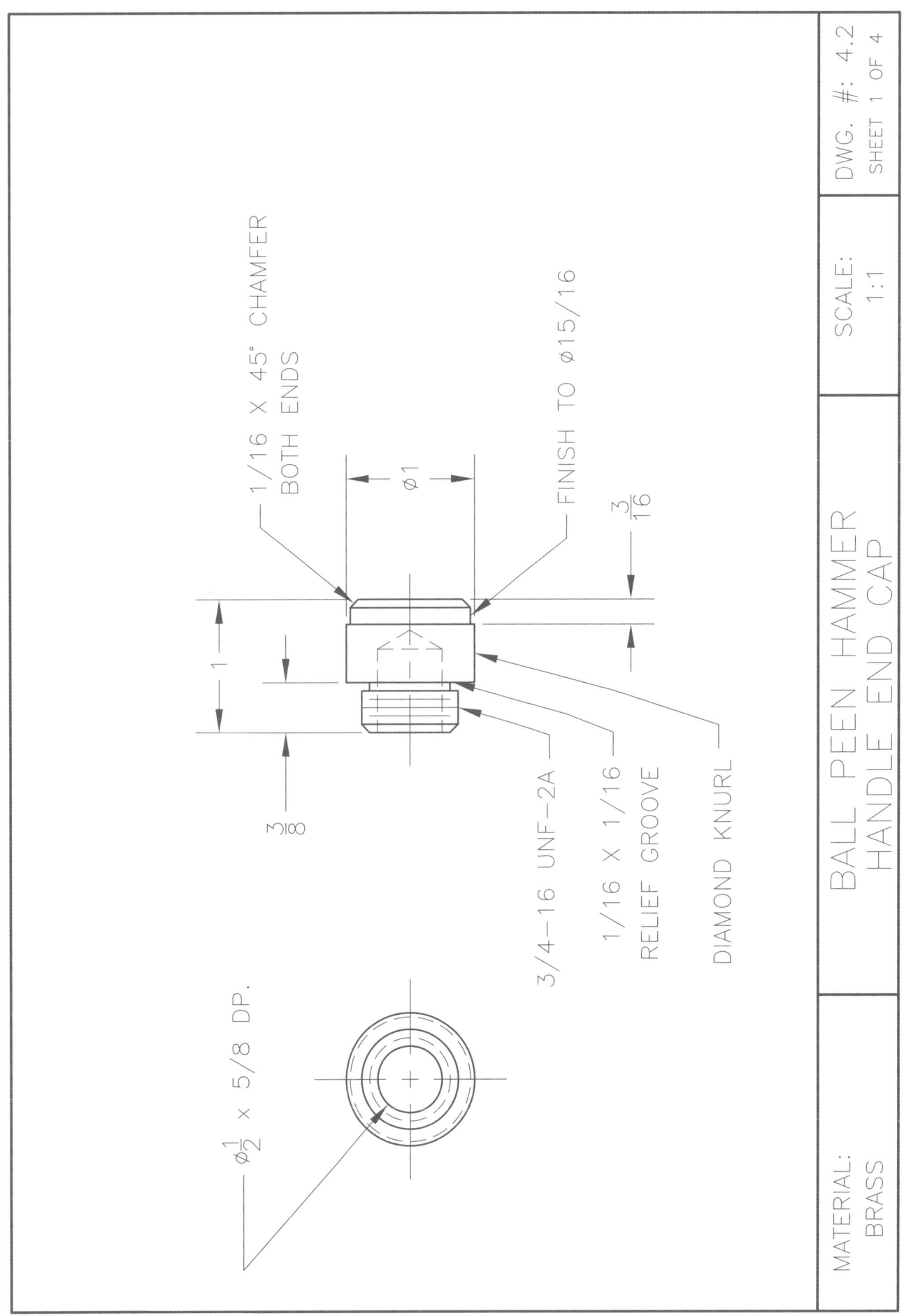

*Machining Projects* **Project 4.2** Ball Peen Hammer

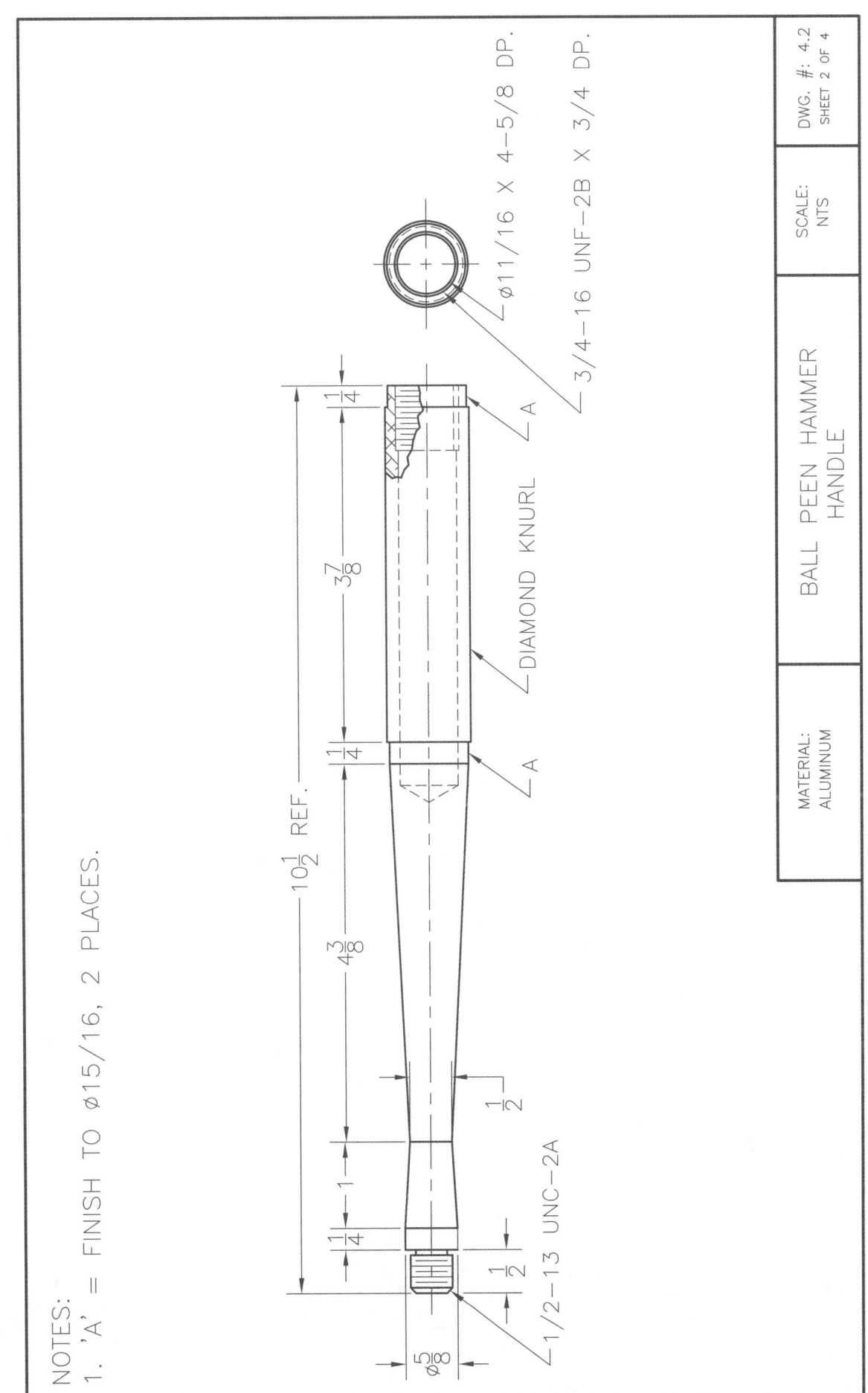

## Project 4.2 Ball Peen Hammer

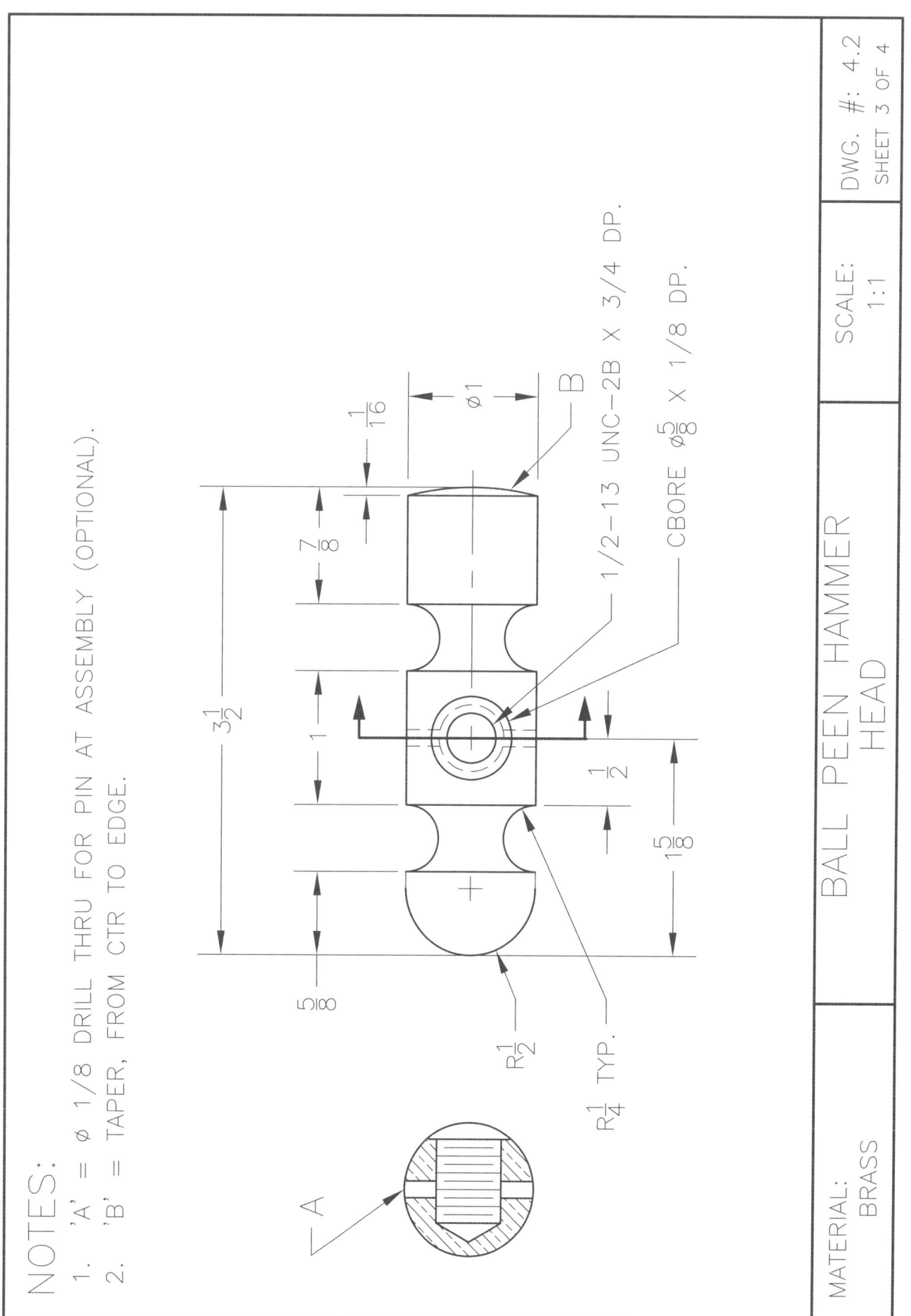

*Machining Projects*  **Project 4.2**  Ball Peen Hammer

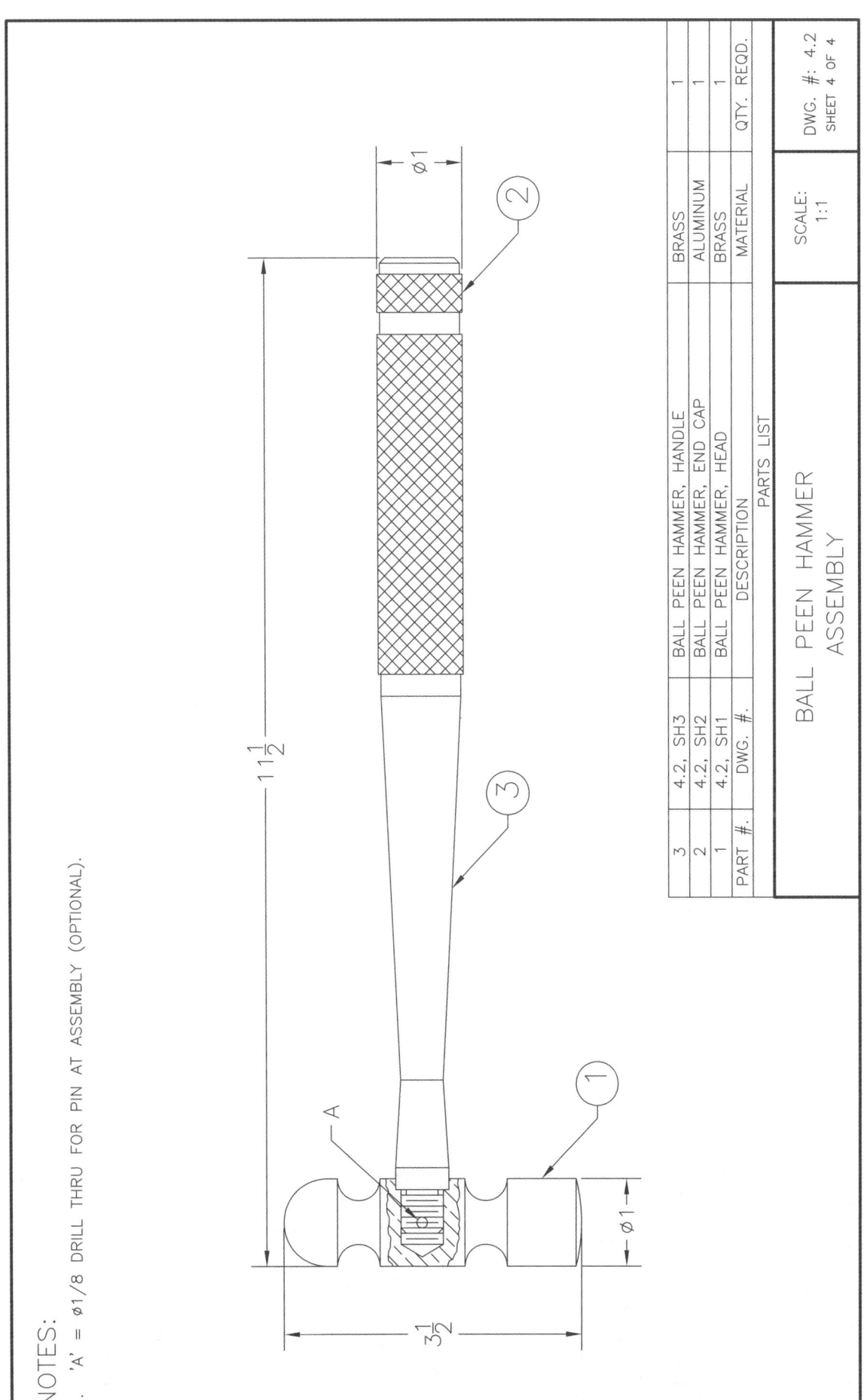

# Project 4.3
# Center Drill Chuck

Name _____ Date _____

Instructor _____ Period ____

## Order of Operations

- [ ] 1. Measure and cut stock plus facing allowance.
- [ ] 2. Face to the specified overall length.
- [ ] 3. Center drill both ends.
- [ ] 4. Drill and ream center hole (for the center drill) to the depth and diameter shown.
- [ ] 5. Locate and cut 1/16" wide × 1/16" deep grooves.
- [ ] 6. Turn the shank portion to the length and diameter specified.
- [ ] 7. Turn the 11/16" diameter portion on the end of the shank.
- [ ] 8. Machine the undercut at the junction of the head and tapered shank.
- [ ] 9. Figure the correct tailstock set-over, taper per inch, taper per foot, or angle of taper that corresponds to a #3 Morse taper and adjust the lathe accordingly.

    Note  Use the textbook or *Machinery's Handbook* for information about the dimensions of Morse tapers.

- [ ] 10. On the vertical mill or drill press, locate, drill, and tap the 1/4-20 setscrew hole.

## Notes

_____
_____
_____
_____
_____
_____
_____
_____
_____
_____

---

**Performance Evaluation—Instructor's Use Only**
Project completed on time?   ☐ Yes   ☐ No
　　If No, which steps were not completed _____
Overall performance on project:
　　☐ Excellent   ☐ Good   ☐ Satisfactory   ☐ Unsatisfactory   ☐ Poor
Comments: _____
_____

Instructor's Signature _____

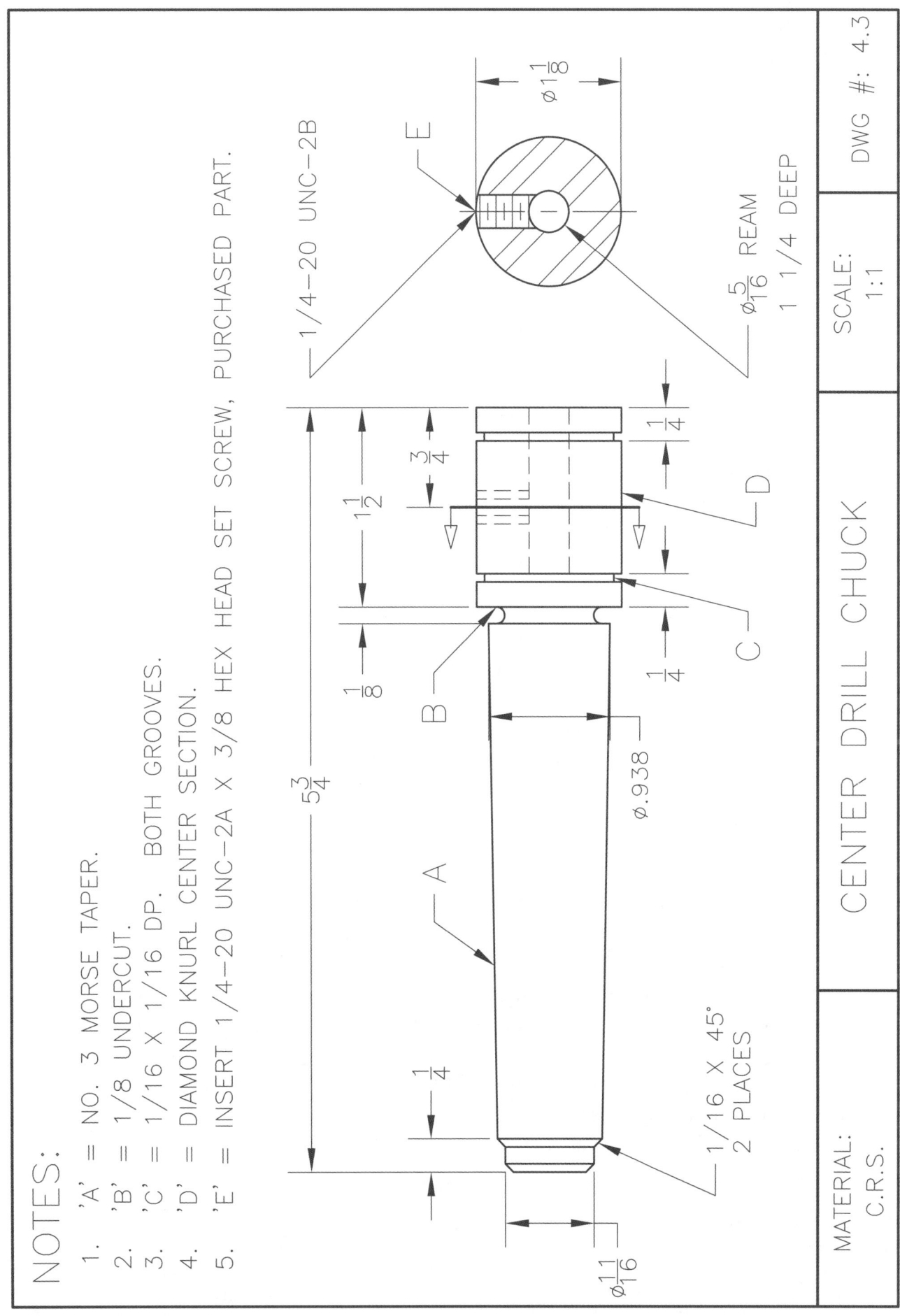

# Project 4.4
## Deburring Tool

Name _____ Date _____

Instructor _____ Period _____

## Order of Operations

- ❏ 1. Add the length of the deburring tool body with the length of the end cap plus 1/4". Cut a piece of 5/8" diameter aluminum to that length.
- ❏ 2. Calculate the correct cutting speed and feed for the type of lathe tool bit being used then set the controls of the lathe accordingly.
- ❏ 3. Face each end just enough to ensure they are straight, smooth, and square.
- ❏ 4. Center drill each end of the workpiece.
- ❏ 5. Drill the body of the workpiece 25/64" × 2 1/2" deep and tap 7/16-20 to the depth indicated.
- ❏ 6. Set up the lathe for knurling. Knurl approximately 3" of the workpiece (some of this will be parted off and used to make the cap).
- ❏ 7. Calculate the correct cutting speed and feed, and then set up the lathe for straight turning.
- ❏ 8. Mount the workpiece either between centers or in a three jaw chuck and turn the knurled end down to 7/16" diameter for a length of 3/8".
- ❏ 9. File or cut a 1/16" × 45° bevel on the end, then thread it 7/16-20.
- ❏ 10. Cut a relief groove to terminate the threads.
- ❏ 11. Part off enough of this end of the workpiece to make the cap. Remember to allow some extra material so the piece can be faced to the correct length for the cap.
- ❏ 12. Face the cap piece to the correct length, then finish the end as shown in the drawing.

    Note    A work-holding device that will prevent having to chuck on the threads can be made by drilling and tapping a 7/16-20 in a piece of round stock which can then be chucked in the lathe with the cap piece screwed into it.

- ❏ 13. Face the knurled end of the remaining workpiece until it is the correct overall length.
- ❏ 14. Drill and ream the knurled end of the workpiece as shown on the drawing.
- ❏ 15. Turn down the steel insert to .002" larger than 3/8" and file a slight bevel on one end.
- ❏ 16. Center drill and drill the end opposite the bevel with a #37 drill × 3/8" deep.
- ❏ 17. Use an arbor press to seat the steel insert flush with the end of the deburring tool body.
- ❏ 18. Place the deburring tool body in the vise of the vertical mill and use an edge finder to accurately position the piece, then drill and tap the 8-32 hole as shown.
- ❏ 19. Turn the appropriate sections of the deburring tool body to .600".
- ❏ 20. Set the compound of the lathe to the specified angle and cut the nose angle as shown.

## Notes

**Performance Evaluation—Instructor's Use Only**
Project completed on time?  ❏ Yes   ❏ No
   If No, which steps were not completed _____
Overall performance on project:
   ❏ Excellent   ❏ Good   ❏ Satisfactory   ❏ Unsatisfactory   ❏ Poor
Comments: _____
_____

Instructor's Signature _____

**Project 4.4** Deburring Tool

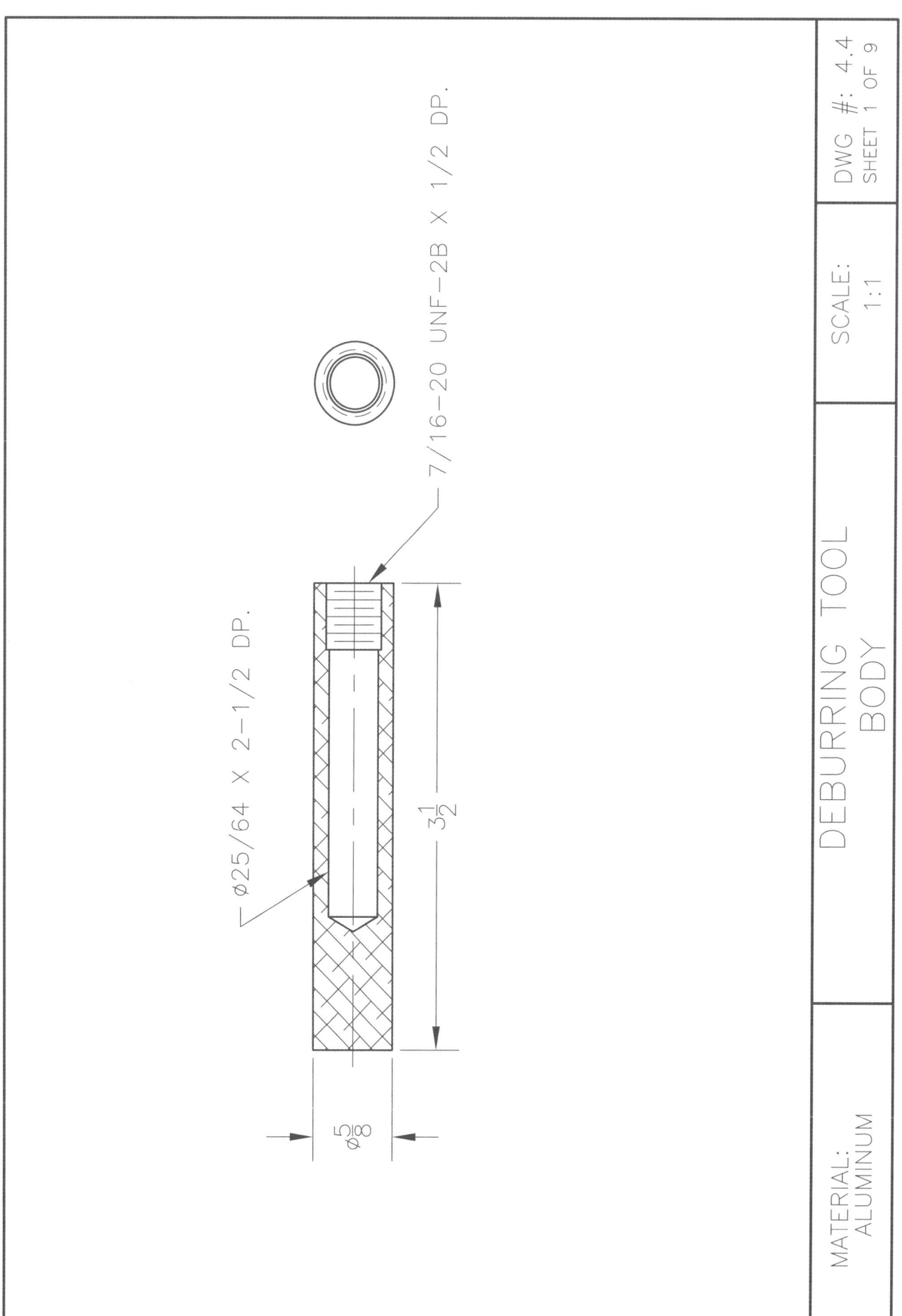

*Machining Projects* — **Project 4.4** Deburring Tool

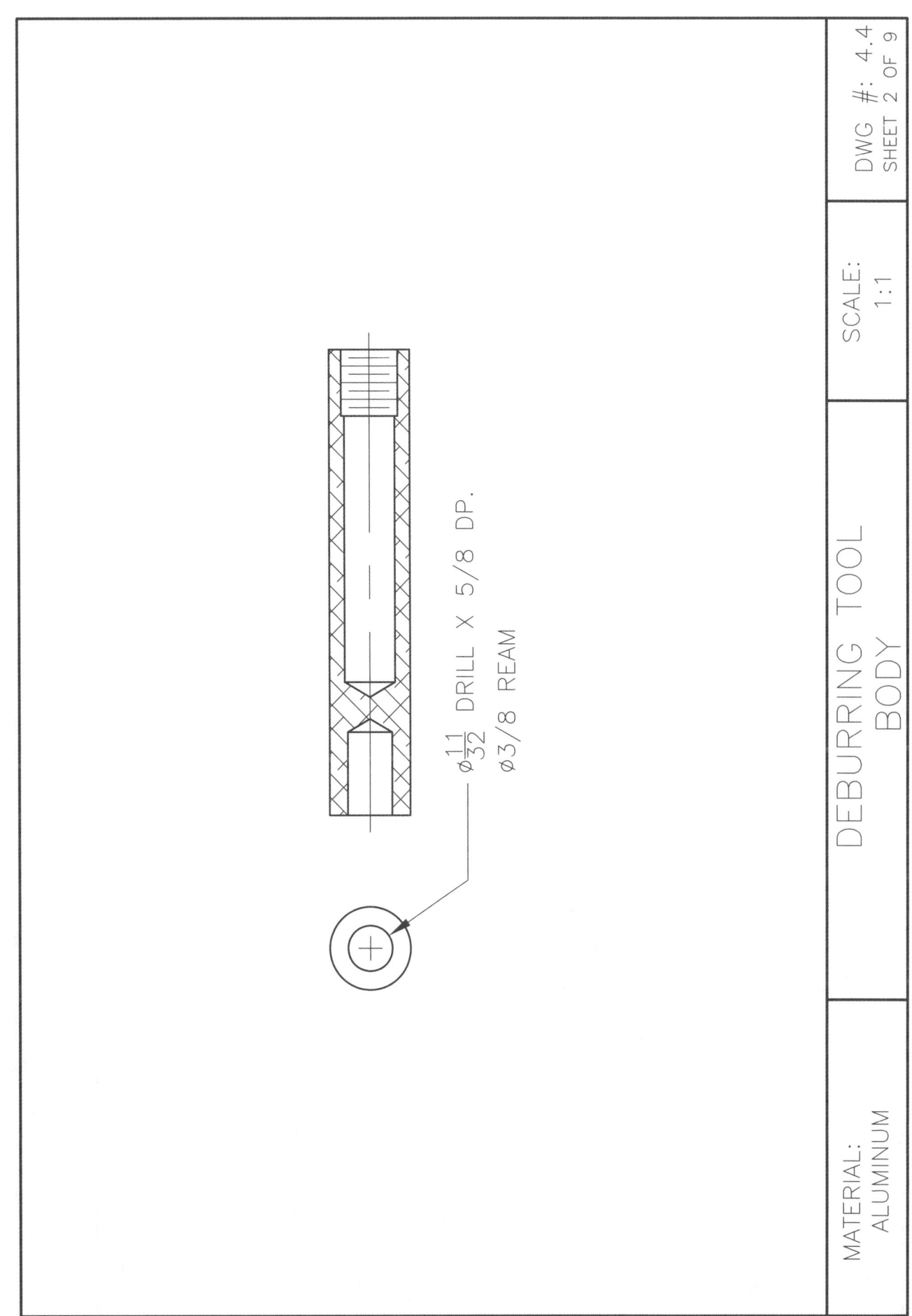

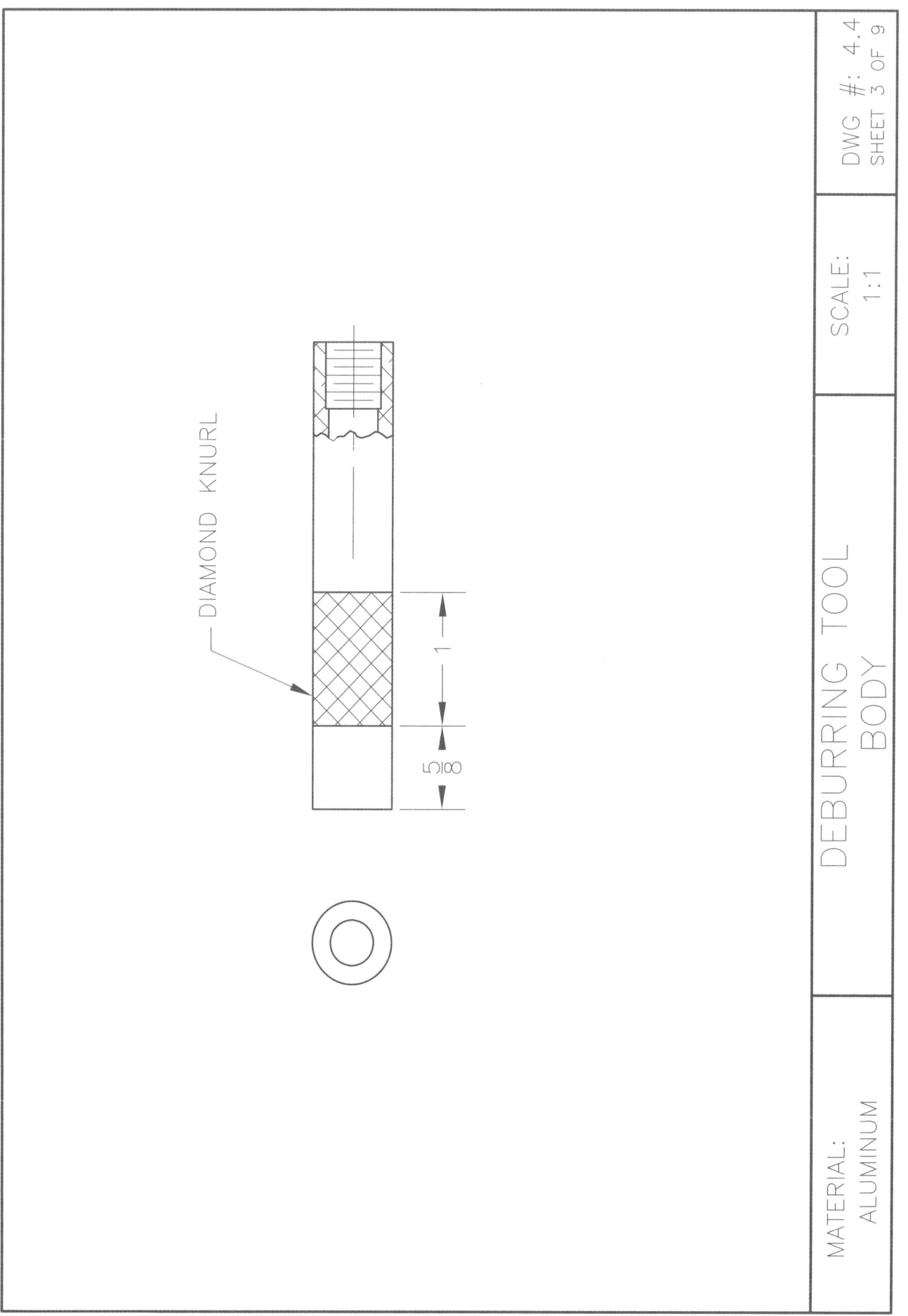

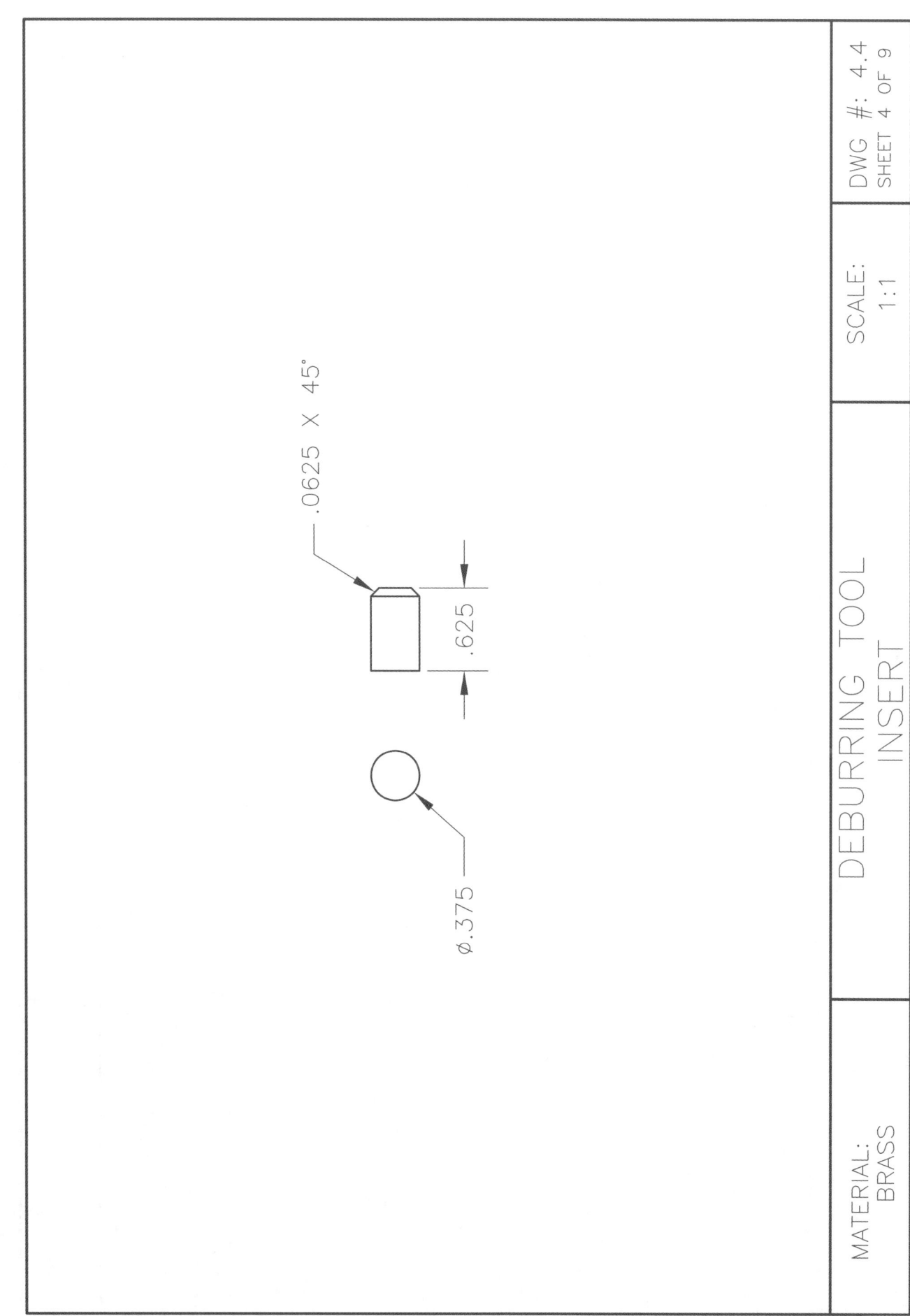

**Project 4.4**  Deburring Tool

NOTES:
1. PRESS FIT PART 2, INSERT, INTO PART 1, BODY.

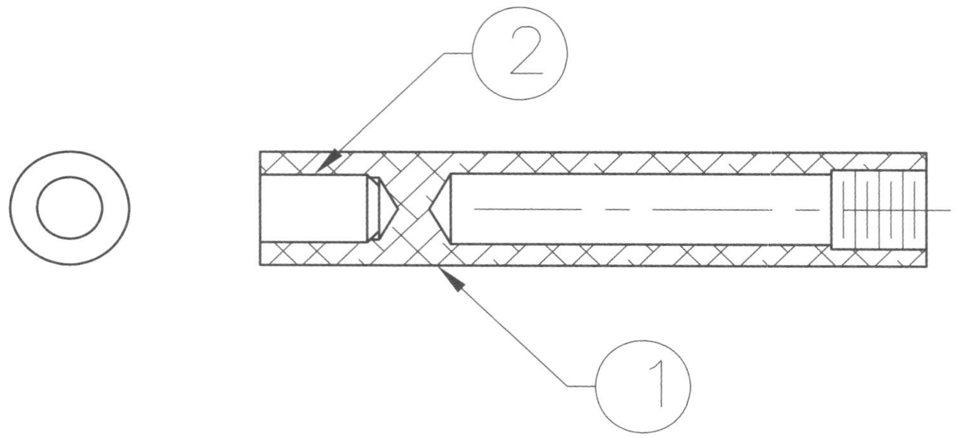

| PART #. | DWG. # | DESCRIPTION | MATERIAL | QTY. REQD. |
|---|---|---|---|---|
| 2 | 4.4, SH4 | DEBURRING TOOL, INSERT | BRASS | 1 |
| 1 | 4.4, SH3 | DEBURRING TOOL, BODY | ALUMINUM | 1 |

PARTS LIST

MATERIAL: ALUMINUM

DEBURRING TOOL SUBASSEMBLY A

SCALE: 1:1

DWG #: 4.4
SHEET 5 OF 9

Machining Projects  Project 4.4  Deburring Tool

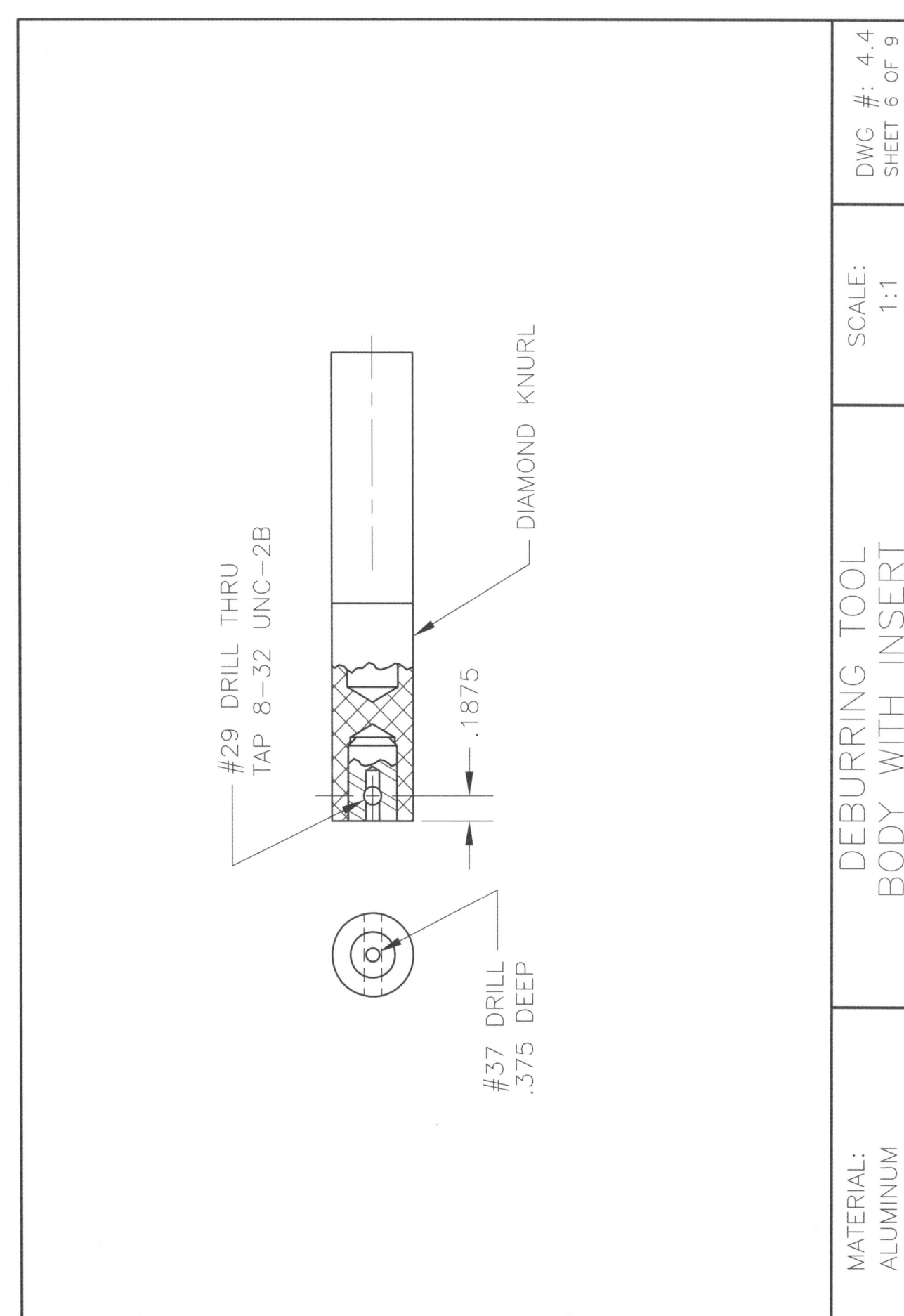

**Project 4.4** Deburring Tool *Machining Projects*

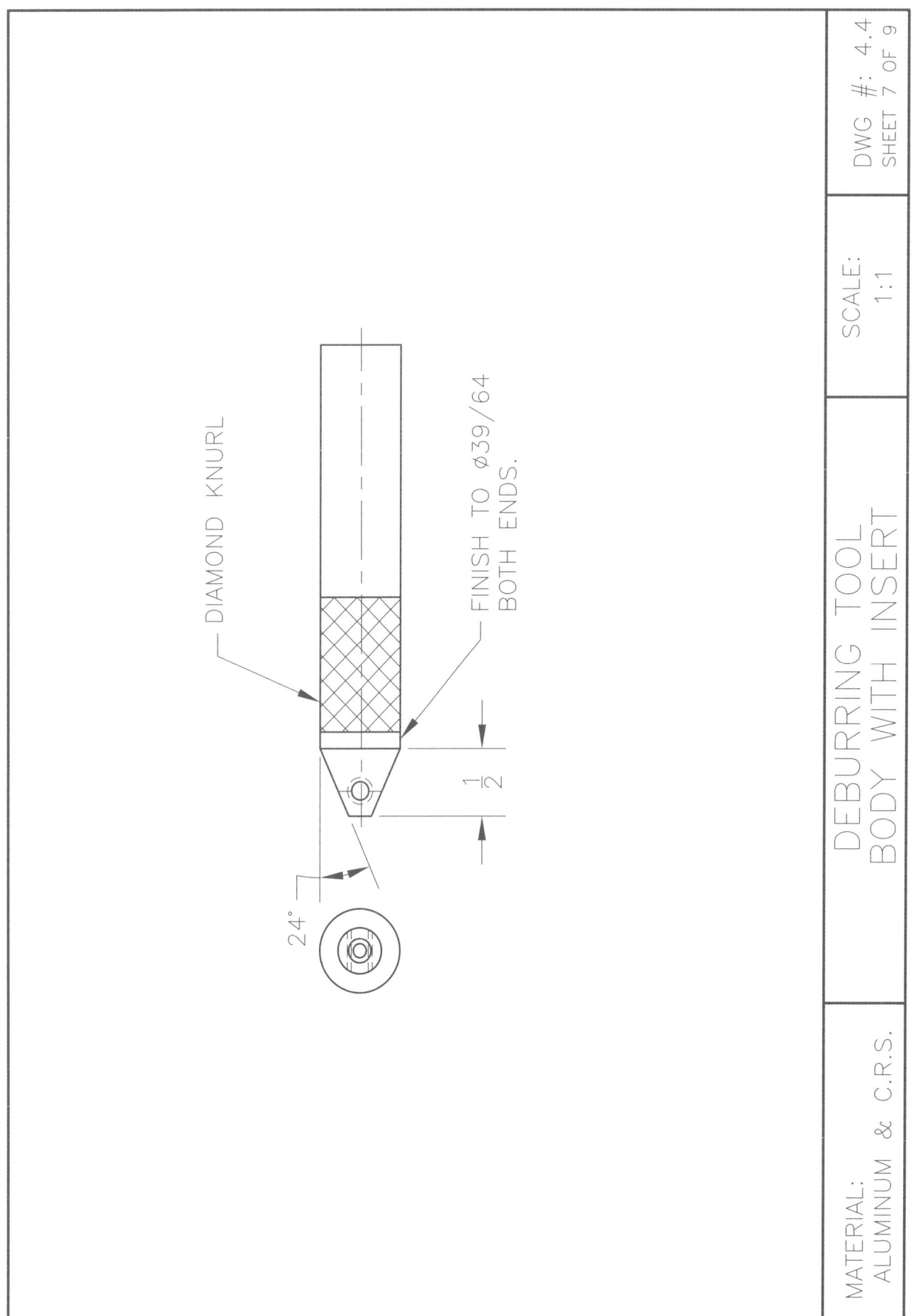

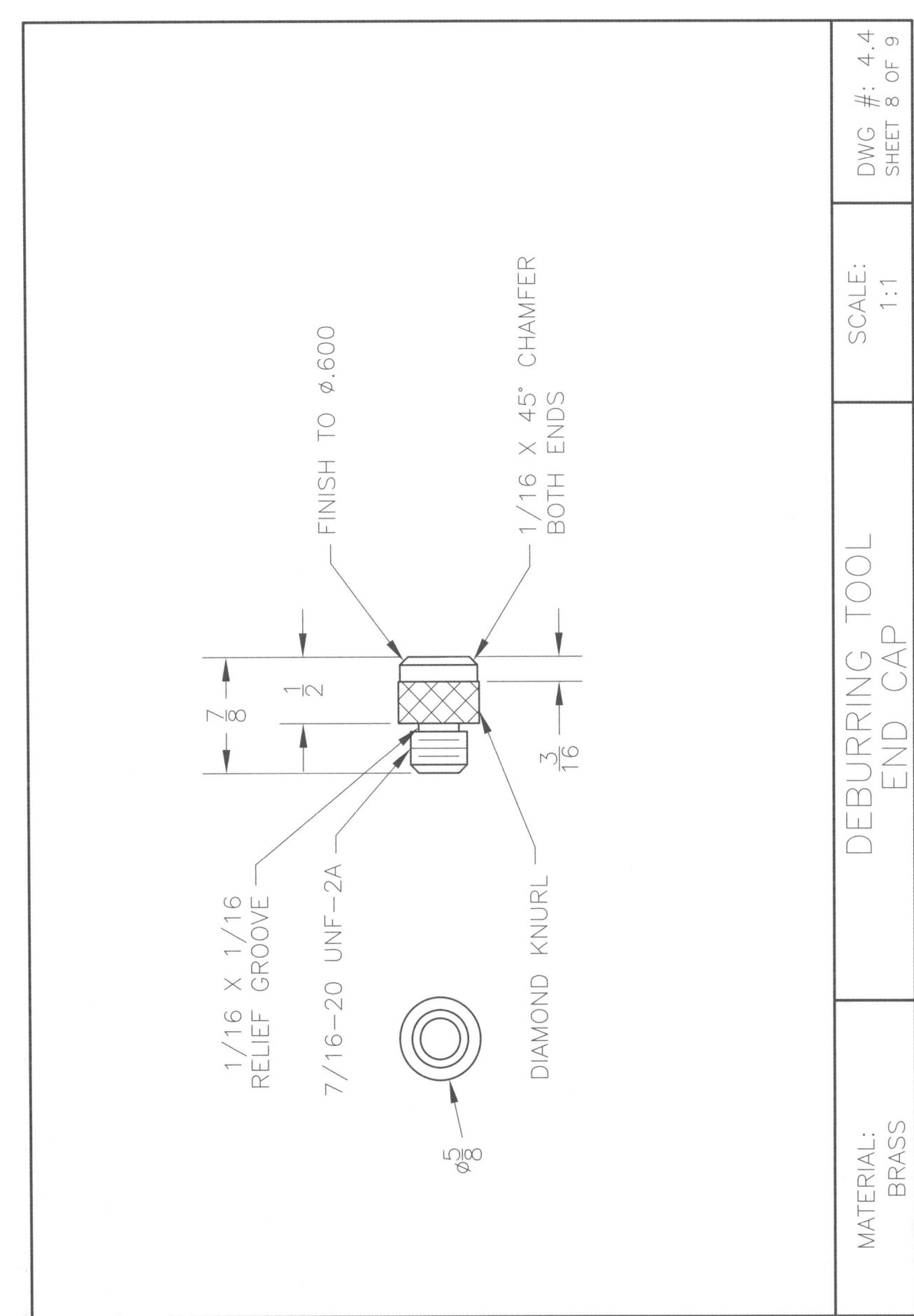

**Project 4.4** Deburring Tool

NOTES:
1. PARTS 4, SET SCREW, AND 5, DEBURRING BLADE, ARE PURCHASED PARTS.

Ø.625

4.000

| PART #. | DWG. #. | DESCRIPTION | MATERIAL: | QTY. REQD. |
|---|---|---|---|---|
| 1 | 4.4, SH3 | BODY | ALUMINUM | 1 |
| 2 | 4.4, SH4 | INSERT | BRASS | 1 |
| 3 | 4.4, SH8 | END CAP | BRASS | 1 |
| 4 | N/A | 8-32 UNC-2A X .125 HEX HD. SET SCREW | STEEL | 1 |
| 5 | N/A | DEBURRING BLADE | STEEL | 1 |

PARTS LIST

DEBURRING TOOL ASSEMBLY

MATERIAL: AS NOTED.

SCALE: 1:1

DWG #: 4.4
SHEET 9 OF 9

# Project 4.5
# Gravity Center Punch

Name _____  Date _____

Instructor _____  Period _____

## Order of Operations

### Hammer
- ❏ 1. Face and center drill one end of a 1″ diameter brass stock.
- ❏ 2. Using the drill bit specified on the print, drill the end of the piece 1/4″ deeper than the overall length of the hammer.
- ❏ 3. Extend the piece approximately 3″ out of the chuck jaws, support the end with a live center mounted in the tailstock spindle.
- ❏ 4. Knurl 1 1/2″ at the end of the piece of stock.
- ❏ 5. Set the compound of the lathe for the angle shown and cut the short taper.
- ❏ 6. Turn down the sections indicated until the knurl is removed (approx. .950″ diameter).
- ❏ 7. Part or cut the workpiece from the bar and face to the overall dimension specified.

### Punch
- ❏ 1. Measure and cut stock plus facing allowance.
- ❏ 2. Face the punch to the correct overall length and center drill one end with a #2 (small) center drill.
- ❏ 3. Turn the section indicated to 1/4″ diameter.
- ❏ 4. Thread the end 1/4-20 × 3/8″ long.
- ❏ 5. Cut a 1/16″ wide × 1/16″ deep groove to cleanly terminate the threads.
- ❏ 6. On the other end of the punch, turn down the end to the length and diameter shown.
- ❏ 7. Set the compound of the lathe for the proper angle and carefully cut the 60° point.
- ❏ 8. Adjust the compound for the 75° angle and cut it.

### Cap
- ❏ 1. Measure and cut stock plus facing allowance.
- ❏ 2. Face the piece to the correct overall length.
- ❏ 3. Drill and tap it 1/4-20 × 3/8″ deep.
- ❏ 4. Cut or file the chamfers as indicated.
- ❏ 5. Sand, polish, and assemble.

    **Note**  The point of the punch should be heat-treated before using.

## Notes

_____

_____

---

**Performance Evaluation—Instructor's Use Only**

Project completed on time?  ❏ Yes   ❏ No

If No, which steps were not completed _____

Overall performance on project:
❏ Excellent  ❏ Good  ❏ Satisfactory  ❏ Unsatisfactory  ❏ Poor

Comments: _____

_____

Instructor's Signature _____

**Project 4.5**  Gravity Center Punch

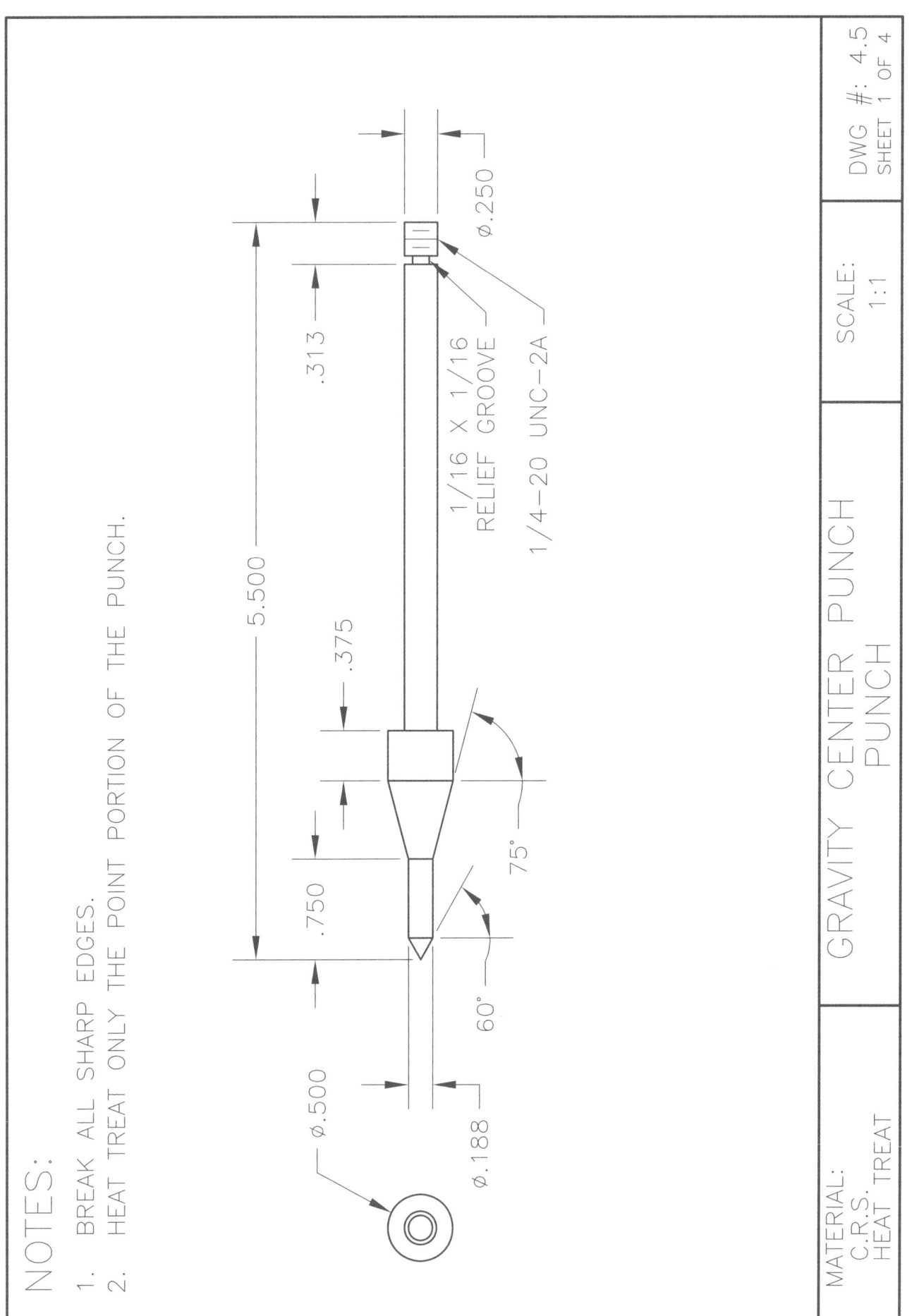

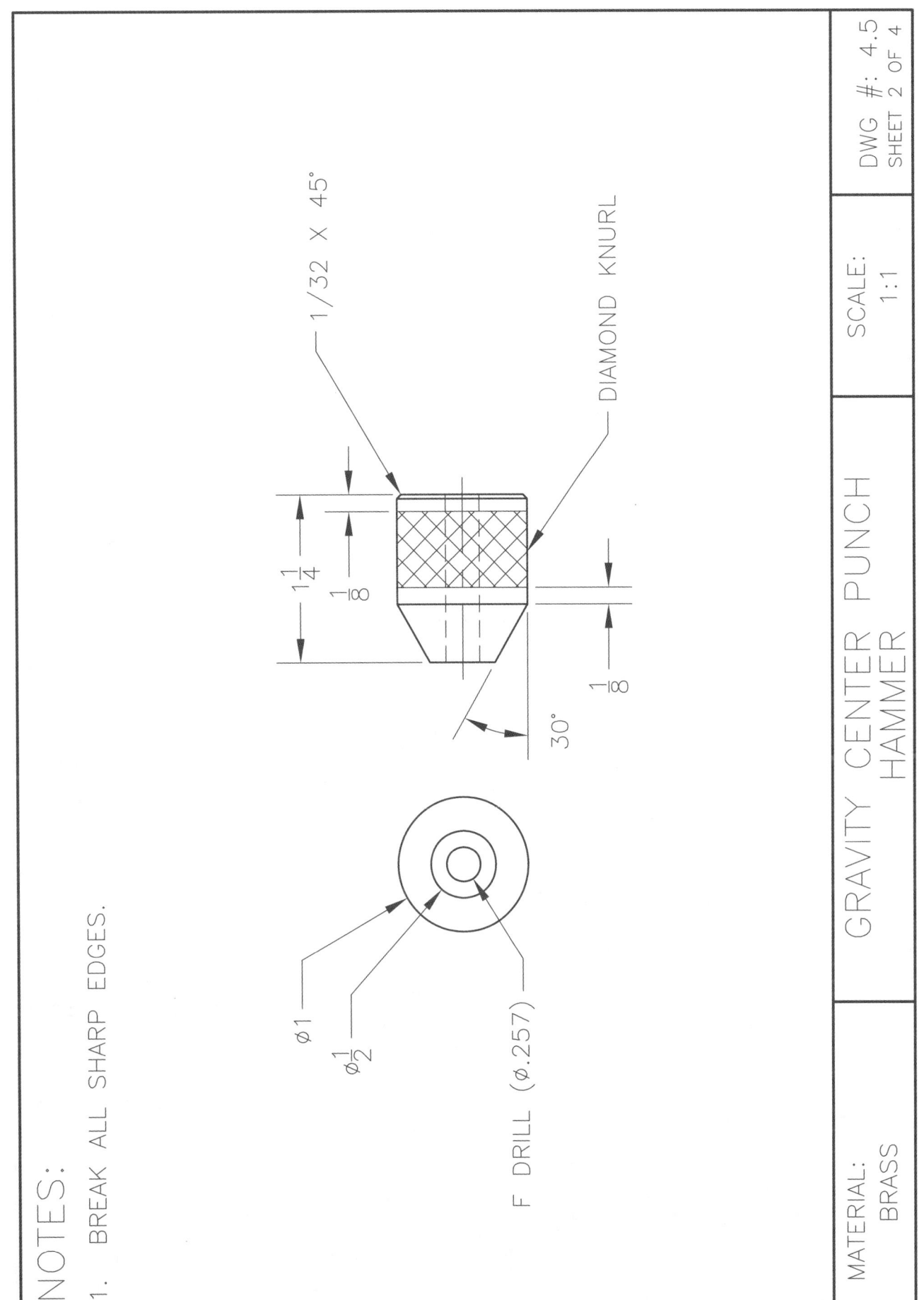

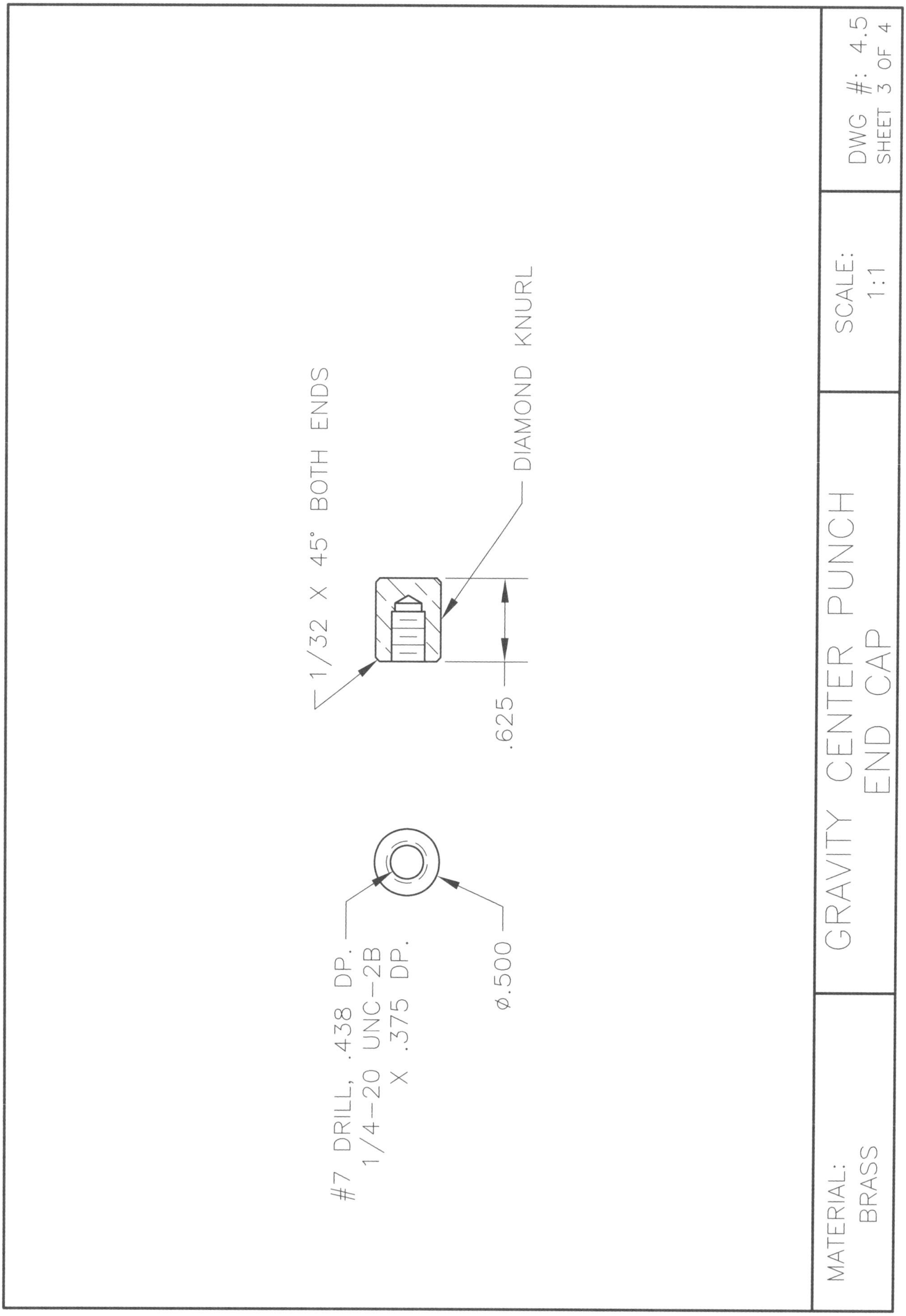

**Project 4.5** Gravity Center Punch

*Machining Projects*

1/32 × 45° BOTH ENDS

DIAMOND KNURL

.625

#7 DRILL, .438 DP.
1/4–20 UNC–2B X .375 DP.

Ø.500

MATERIAL: BRASS

GRAVITY CENTER PUNCH
END CAP

SCALE: 1:1

DWG #: 4.5
SHEET 3 OF 4

*Machining Projects*      **Project 4.5**      Gravity Center Punch

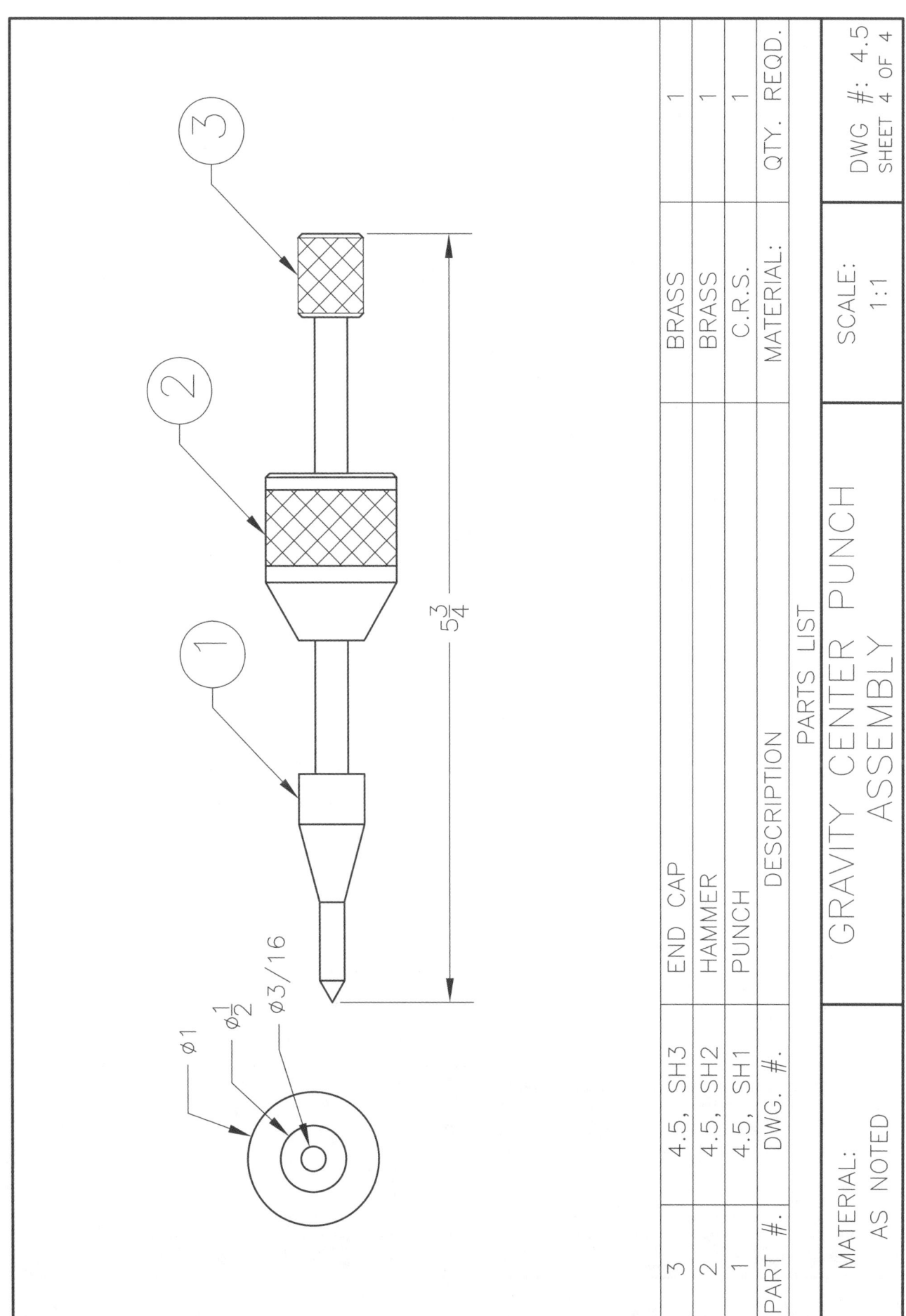

# Project 4.6
## Fly Cutter

Name _____  Date _____

Instructor _____  Period _____

## Order of Operations

☐ 1. Measure and cut stock plus facing allowance.
☐ 2. Face to the specified length.
☐ 3. Turn the shank section to the length and diameter indicated.
☐ 4. Figure the angle of the cutter head face, then set the mill vise accordingly.
☐ 5. Place the fly cutter in the mill vise in a horizontal position and cut the face angle as shown.
☐ 6. Using an arbor-mounted plain milling cutter 5/16" wide, cut the tool bit slot to the depth specified on the print.
☐ 7. Cut the other step indicated using the same set up.
☐ 8. Mount a drill chuck in the mill and locate, drill, countersink, and tap the three setscrew holes.
☐ 9. Realign the mill vise (always check with a dial indicator) and locate and cut the slot on the side for initials.
☐ 10. Stamp maker's initials, deburr, and polish as desired.

## Notes

_____
_____
_____
_____
_____
_____
_____
_____
_____
_____

---

**Performance Evaluation—Instructor's Use Only**
Project completed on time?  ☐ Yes  ☐ No
   If No, which steps were not completed _____
Overall performance on project:
   ☐ Excellent  ☐ Good  ☐ Satisfactory  ☐ Unsatisfactory  ☐ Poor
Comments: _____
_____

Instructor's Signature _____

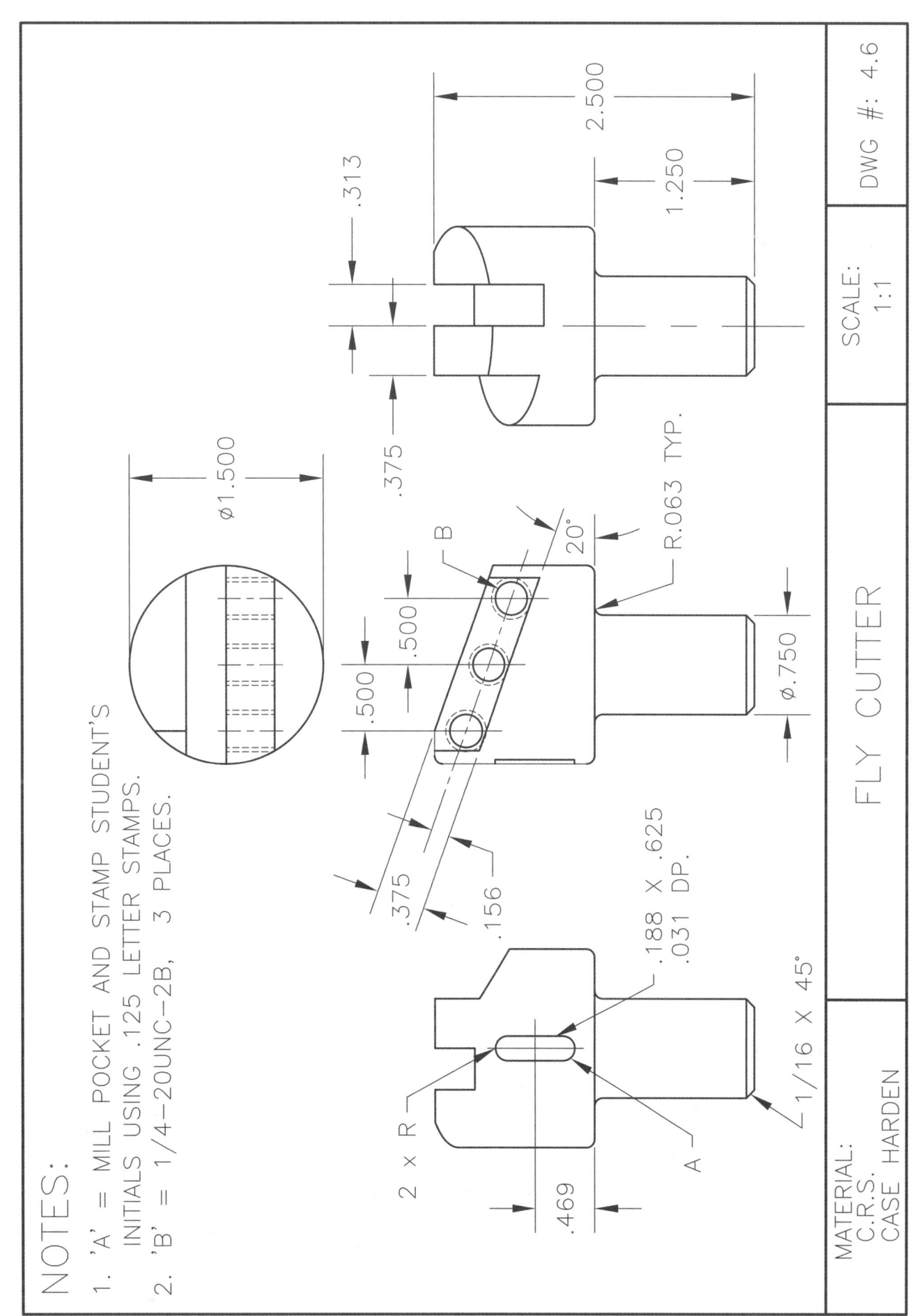

# Project 4.7
## Lathe Puzzle

Name _____ Date _____

Instructor _____ Period _____

## Order of Operations

### Left End
- ❏ 1. Measure and cut stock plus facing allowance.
- ❏ 2. Face to the specified length.
- ❏ 3. Center drill, drill, countersink, and tap one end 3/8-16 to a depth of 1 1/4".
- ❏ 4. Turn the portion to be threaded to .750" diameter.
- ❏ 5. Cut the relief groove as indicated on the print.

### Right End
- ❏ 1. Measure and cut stock plus facing allowance.
- ❏ 2. Center drill the end that will become the threaded end.
- ❏ 3. Turn the end to be threaded to .750".
- ❏ 4. Turn 1" of that end to 3/8" and thread 3/8-16.
- ❏ 5. Cut the relief groove as indicated on the print.

### Center Collar
- ❏ 1. Cut and face a piece of stock of the diameter specified.
- ❏ 2. Center drill, drill, countersink, and tap 3/4-16 for at least 1".
- ❏ 3. Knurl at least 1 1/2" of the piece.
- ❏ 4. Part off the section that will become the nut.

    Note    Remember, you should never attempt parting operations on a piece that is held between centers or supported by a live center!

- ❏ 5. Face the nut (both sides) to the specified dimension.

    Note    To avoid damaging the knurl, a piece of 3/4-16 threaded rod or any piece that has been threaded to 3/4-16 can be placed in the lathe chuck and the nut mounted to it for facing.

### Threaded Bolt Assembly
- ❏ 1. Assemble right end and left end (the 3/8-16 section should be lubricated or be covered with a thin coat of antiseize compound to prevent galling).
- ❏ 2. Set up the lathe for chasing 16 threads per inch.
- ❏ 3. Chase the threads to the appropriate depth. Check thread depth frequently to achieve the best possible fit and finish.
- ❏ 4. Separate the two sections, install the nut, and reassemble.

**Notes**

---

**Performance Evaluation—Instructor's Use Only**
Project completed on time?  ❏ Yes  ❏ No
    If No, which steps were not completed _____
Overall performance on project:
    ❏ Excellent  ❏ Good  ❏ Satisfactory  ❏ Unsatisfactory  ❏ Poor
Comments: _____

Instructor's Signature _____

**Project 4.7** Lathe Puzzle  *Machining Projects*

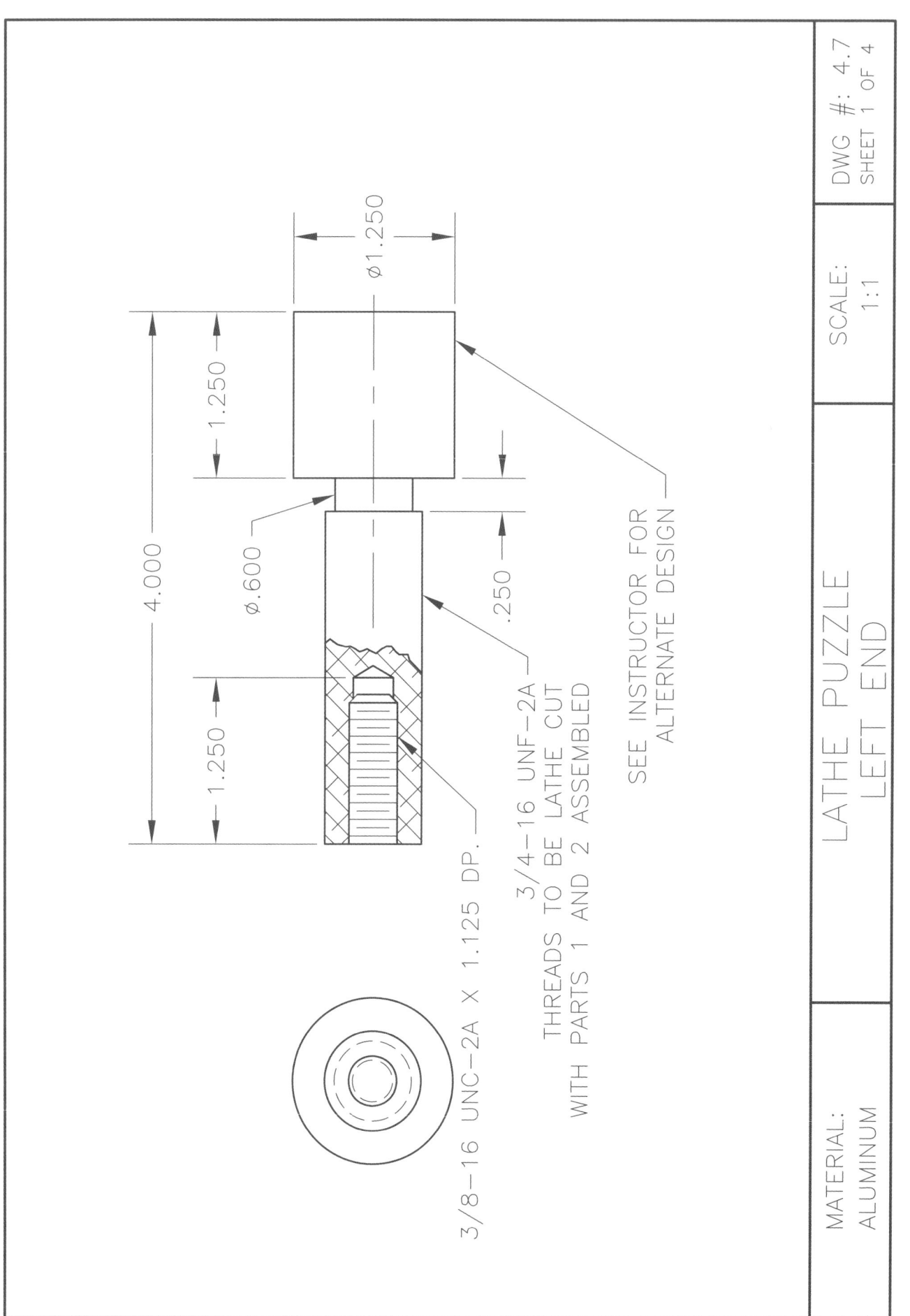

LATHE PUZZLE
LEFT END

DWG #: 4.7
SHEET 1 OF 4

SCALE: 1:1

MATERIAL: ALUMINUM

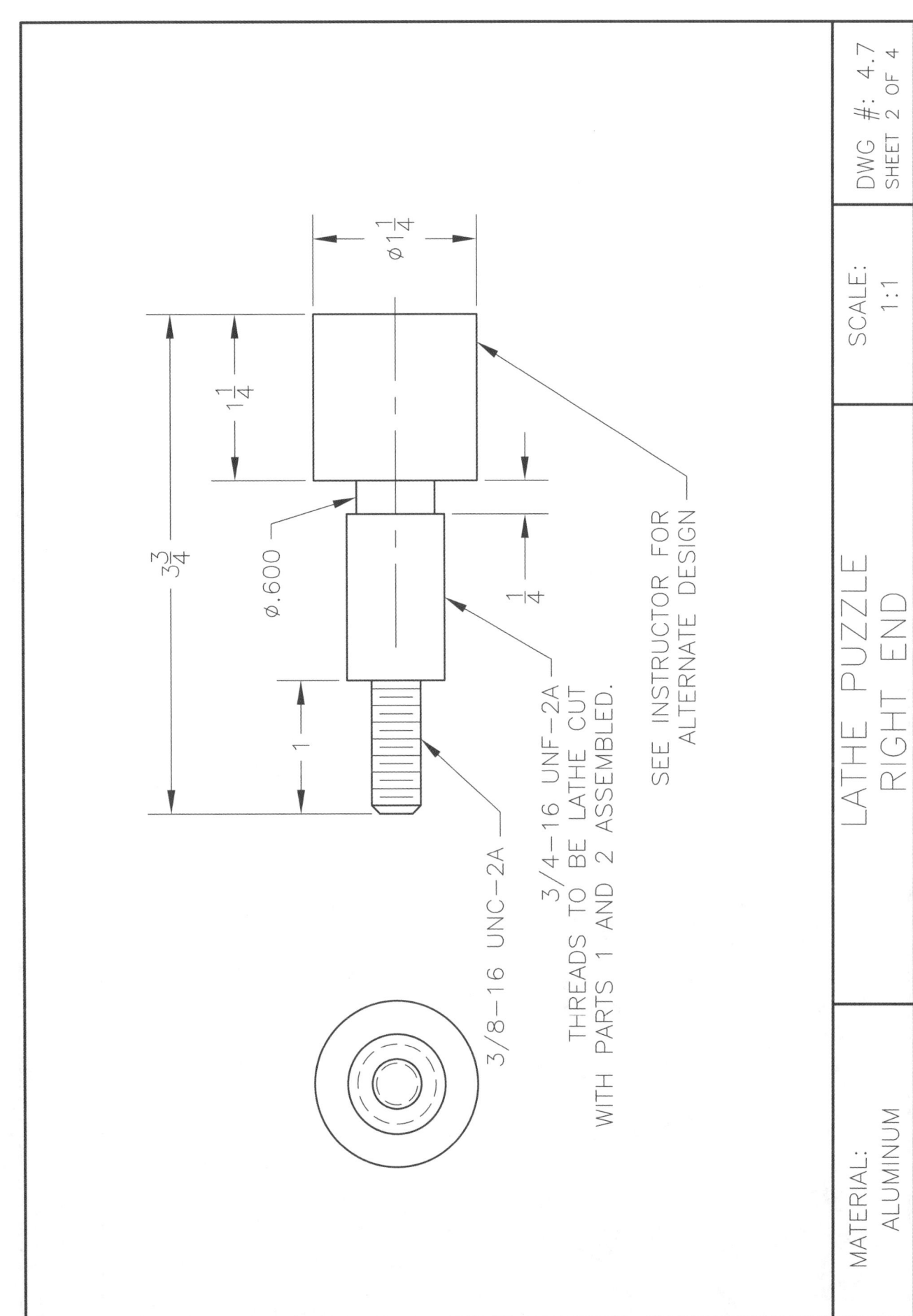

**Project 4.7** Lathe Puzzle

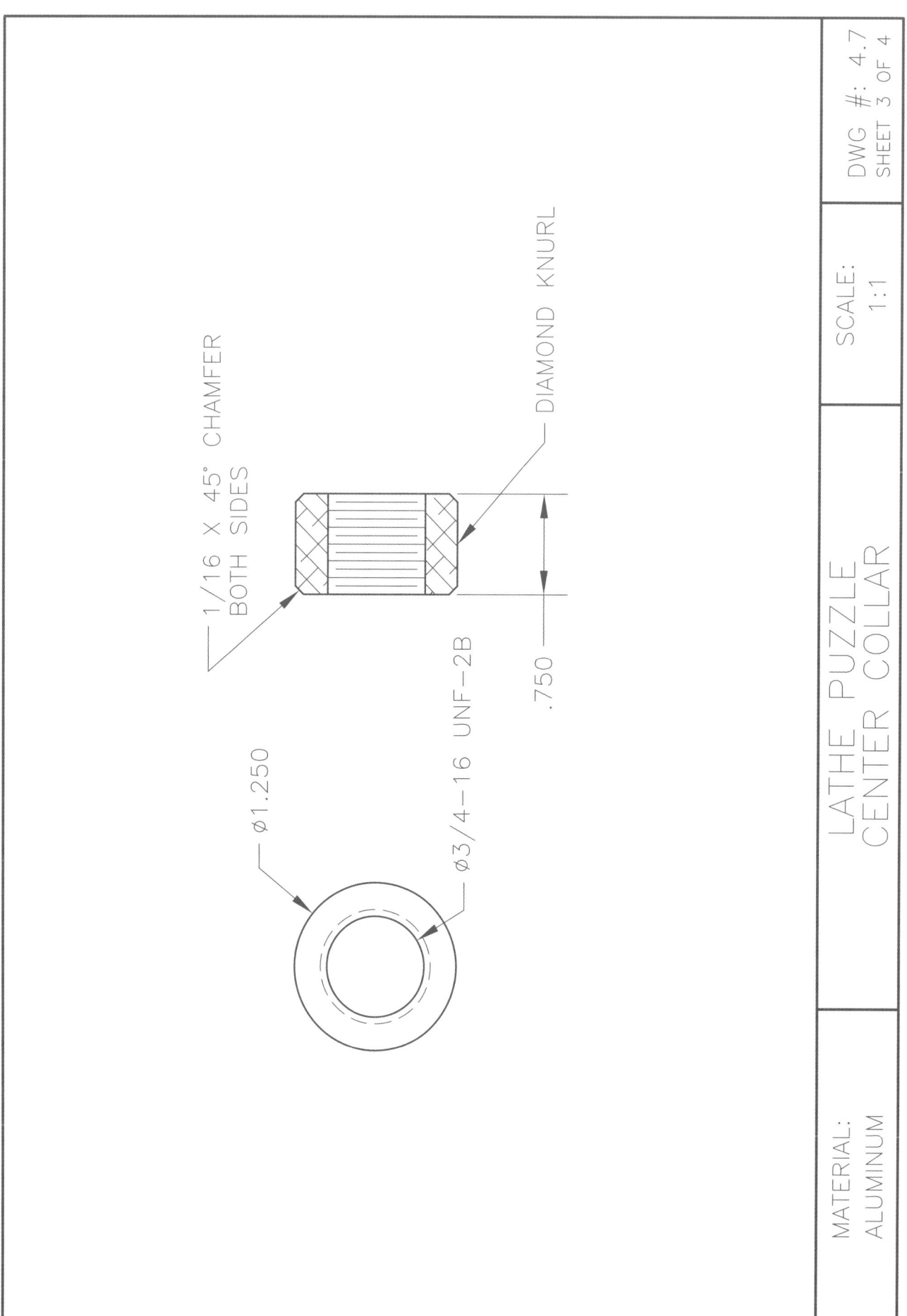

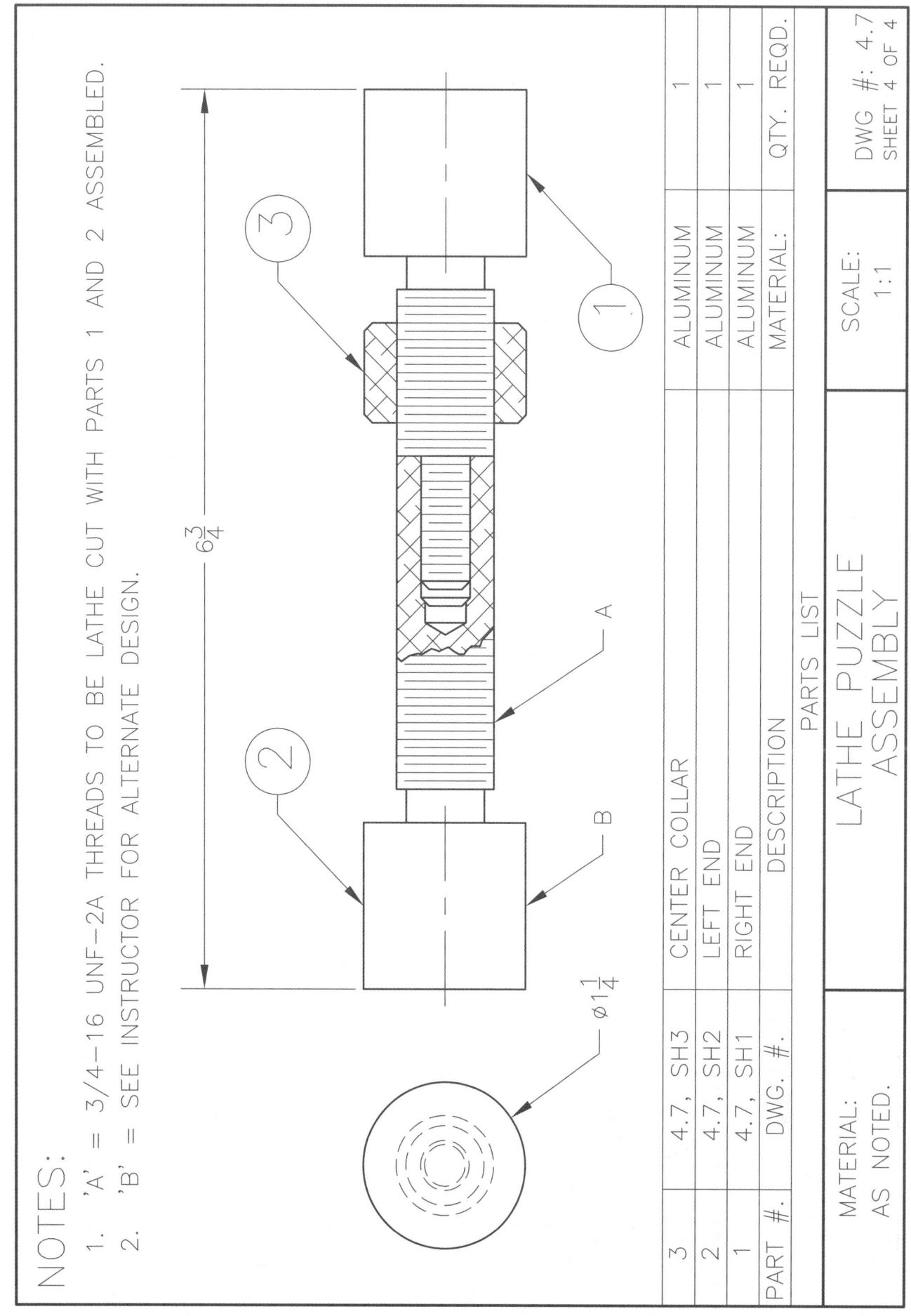

Project 4.8   Mill Cutter Arbor                                                          *Machining Projects*

# Project 4.8
# Mill Cutter Arbor

Name _____  Date _____

Instructor _____  Period _____

## Order of Operations

- ❏ 1. Cut appropriate stock 1/8″ longer than the finished length specified on the drawing.
- ❏ 2. Calculate the correct cutting speed and feed for the type of tool bit and material being used.
- ❏ 3. Face the workpiece to the specified overall length.
- ❏ 4. Center drill both ends of the workpiece.
- ❏ 5. Set up workpiece between centers and cut it to the specified dimensions.
- ❏ 6. Set the compound for 35° and cut the angle.
- ❏ 7. Set up the lathe for cutting threads and cut the 5/8-18 thread.
- ❏ 8. Check the alignment of the vertical mill head with a dial indicator and trim if necessary.
- ❏ 9. Check the alignment of the mill vise with a dial indicator and adjust if necessary.
- ❏ 10. Calculate the correct cutting speed and feed for the type of end mill being used.
- ❏ 11. Cut the keyway as indicated.
- ❏ 12. Deburr all edges.

## Notes

_____
_____
_____
_____
_____
_____
_____
_____
_____

---

**Performance Evaluation—Instructor's Use Only**

Project completed on time?   ❏ Yes   ❏ No

  If No, which steps were not completed _____

Overall performance on project:

  ❏ Excellent   ❏ Good   ❏ Satisfactory   ❏ Unsatisfactory   ❏ Poor

Comments: _____
_____

Instructor's Signature _____

Machining Projects

**Project 4.8**  Mill Cutter Arbor

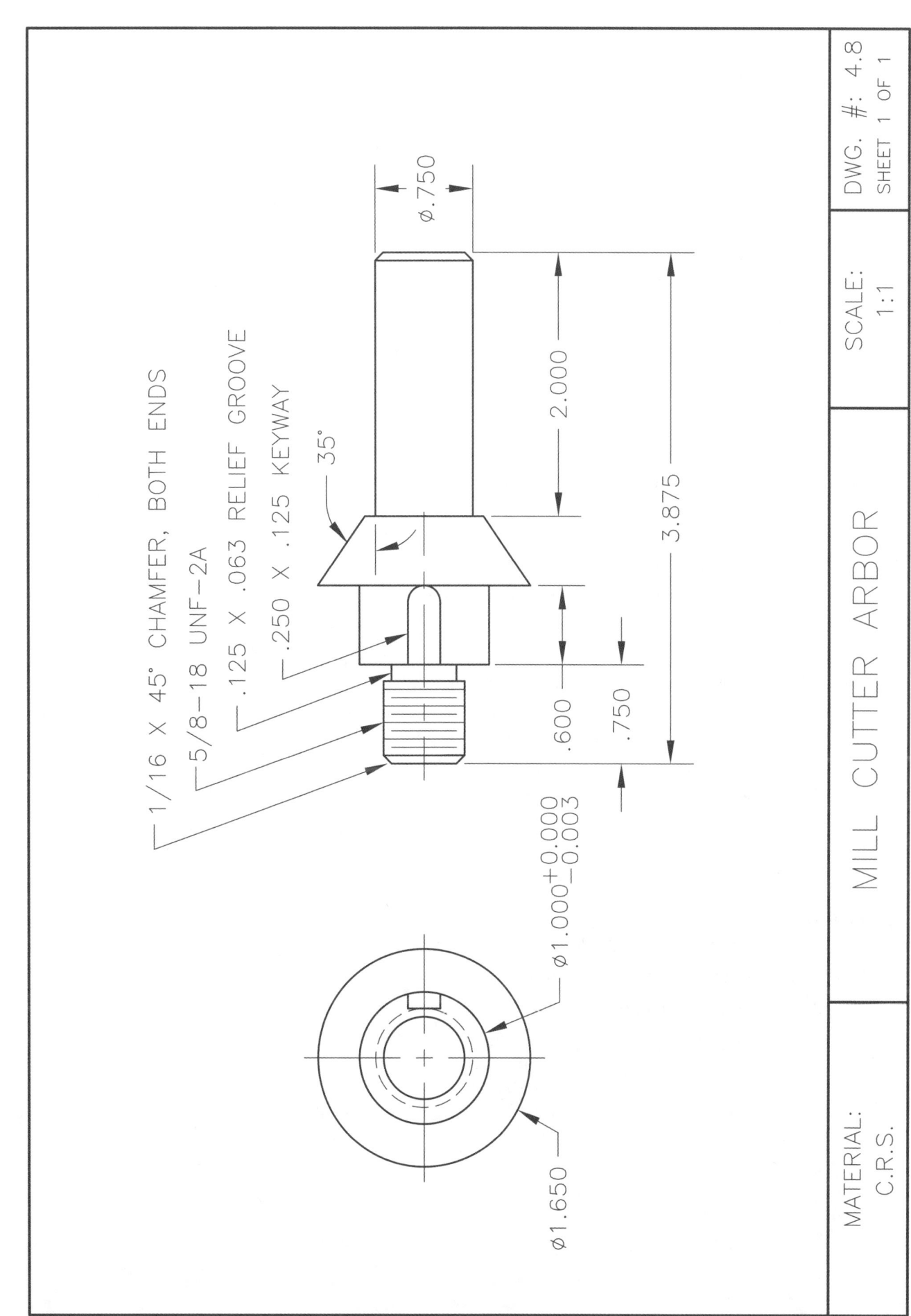

# Project 4.9
# Model Civil War Cannon

Name _____  Date _____

Instructor _____  Period _____

## Order of Operations

**Barrel**

❏ 1. Cut stock plus facing allowance.
❏ 2. Calculate the correct cutting speed and feed for the material and type of tool bit to be used.
❏ 3. Face workpiece to the specified length.
❏ 4. Center drill one end.
❏ 5. Drill and ream the bore.
❏ 6. Set up the lathe for straight turning and turn the workpiece to the largest diameter shown on the print.
❏ 7. Turn the section of the breech end that will become the acorn finial to the specified diameter.
❏ 8. Cut the undercut and radius on the breech end.
❏ 9. Turn the barrel section to 7/8" for the length indicated on the print.
❏ 10. Set up the vertical mill (or drill press) for drilling and reaming the cross-pin hole. If using a vertical mill, be sure to check the trim and vise alignment with a dial indicator.
❏ 11. Locate the hole position and drill and ream.
❏ 12. Using the *Machining Fundamentals* textbook or *Machinery's Handbook* as a reference, figure the tailstock setover, the taper per inch, the taper per foot, or the degree of taper. (Use the one that is applicable to the equipment available.)
❏ 13. Set the tailstock or taper attachment to the correct position and cut the tapered portion of the cannon barrel.
❏ 14. Turn the workpiece around, secure in the chuck, and file the acorn-shaped finial on the breech. A tool bit ground to the appropriate shape can also be used.

**Cross-Pin**

❏ 1. Measure and cut stock plus facing allowance.
❏ 2. Face to the correct overall length.
❏ 3. Turn down each end to the dimensions specified.
❏ 4. Using a file, slightly bevel both ends of the cross-pin.

## Barrel Support Mounts—Two Required

- ❏ 1. Measure and cut stock plus facing allowance.
- ❏ 2. Set up the vertical mill and face the pieces to the specified dimensions.
- ❏ 3. Calculate the correct angles for the support mounts, and then adjust the mill vise accordingly. Remember that there is a right-hand mount and left-hand mount and they are *not* exactly the same.
- ❏ 4. Mill the angles on the support mounts.
- ❏ 5. When the above operations have been performed, move the mill vise back to its original zero and check with a dial indicator.
- ❏ 6. Locate the hole centers for the cross-pin holes and drill and ream. Use of an edge finder is recommended.
- ❏ 7. Locate and drill the holes for the base mounting screws.
- ❏ 8. Deburr and break all holes and sharp edges.

## Base

- ❏ 1. Measure and cut stock plus facing allowance for the cannon base.
- ❏ 2. Cut the 45° angle on the top side of the cannon base.

  Note   Use of a ball nose end mill centered on the edges and down-fed half the thickness of the base will result in a very attractive radius.

- ❏ 3. Locate, drill, countersink, and tap the four holes for the mounting screws.
- ❏ 4. Deburr sharp edges, sand or polish, and assemble.

## Notes

_____
_____
_____
_____
_____
_____
_____
_____
_____
_____
_____

---

**Performance Evaluation—Instructor's Use Only**

Project completed on time?    ❏ Yes    ❏ No
    If No, which steps were not completed _____
Overall performance on project:
    ❏ Excellent    ❏ Good    ❏ Satisfactory    ❏ Unsatisfactory    ❏ Poor
Comments: _____
_____

Instructor's Signature _____

**Project 4.9** Model Civil War Cannon

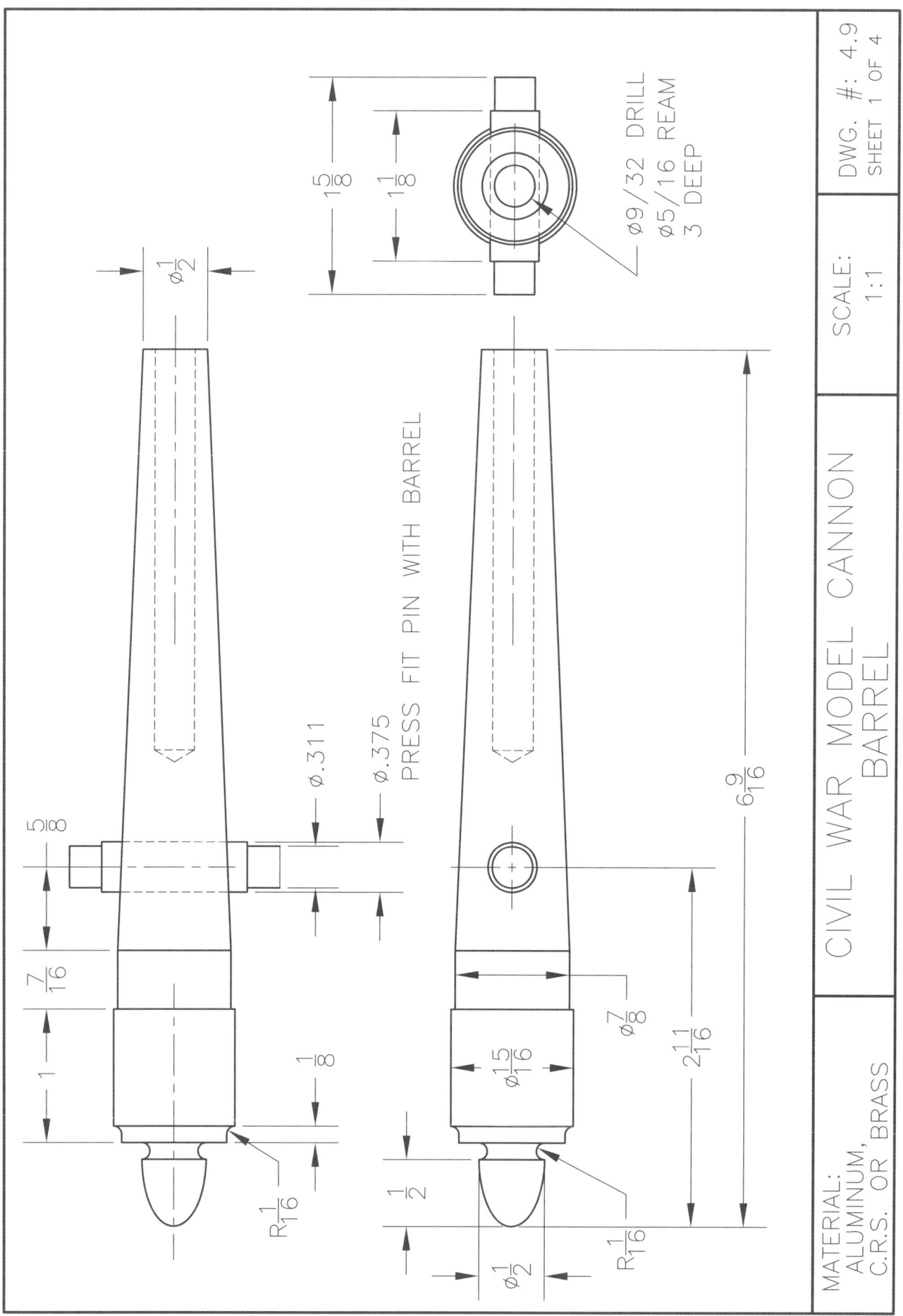

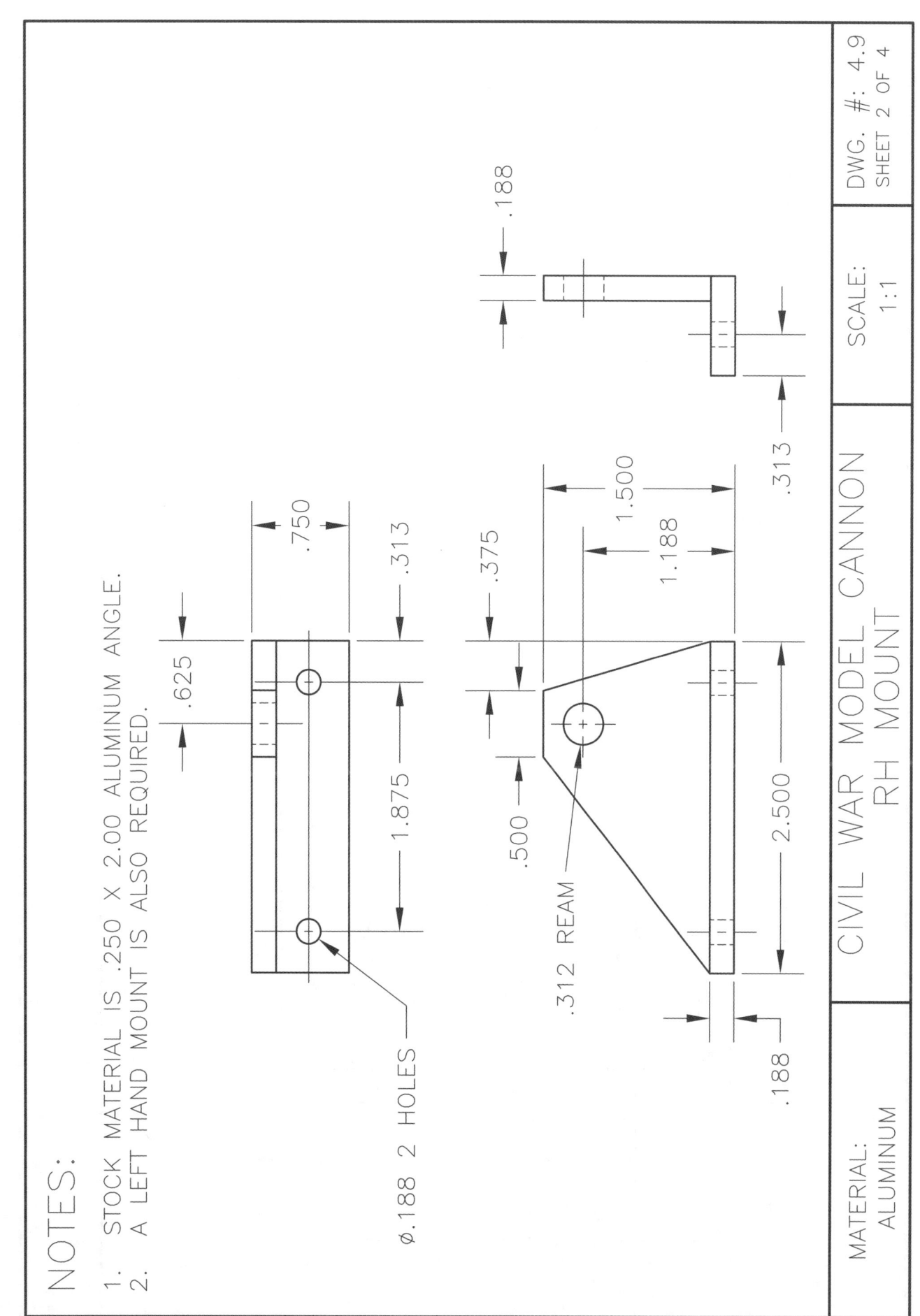

**Project 4.9** Model Civil War Cannon

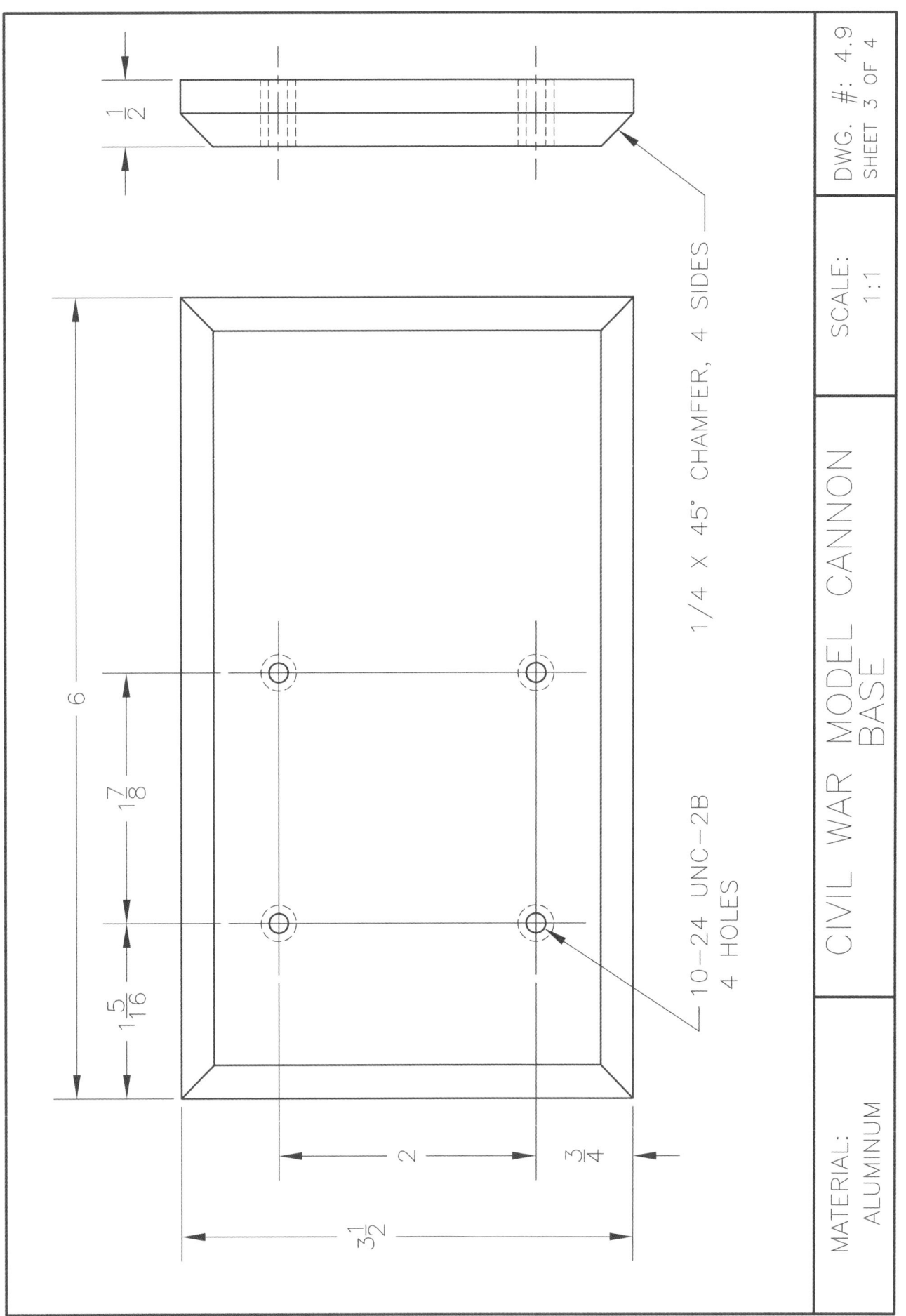

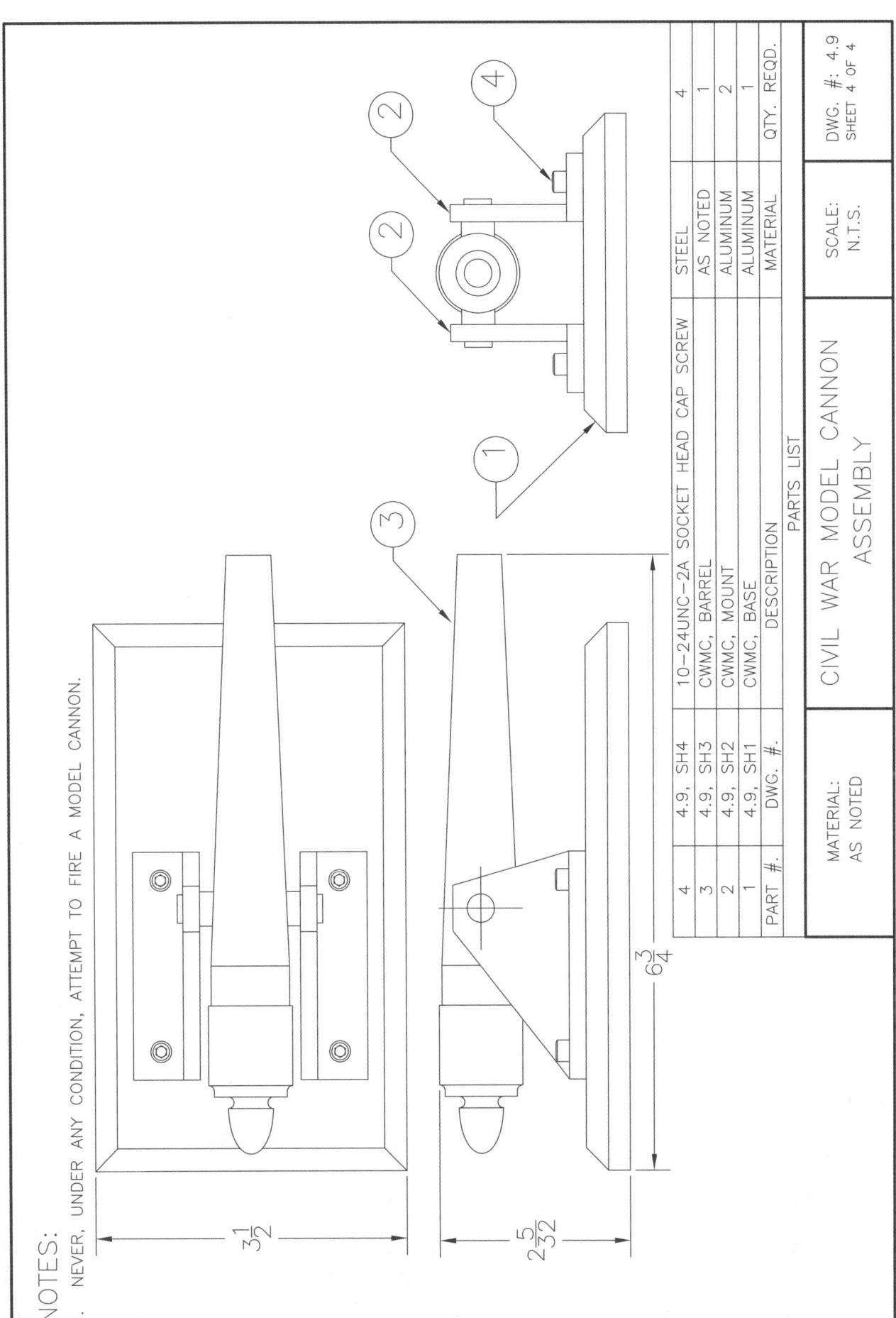

# Project 4.10
# R-8 Tool Holder

Name _____ Date _____

Instructor _____ Period _____

## Order of Operations

- ❏ 1. Measure and cut 2" diameter stock length plus facing allowance.
- ❏ 2. Face workpiece to an overall length of 6.125".
- ❏ 3. Center drill both ends.
- ❏ 4. Set up the lathe for turning between centers.
- ❏ 5. Turn the shank to .950" diameter × 3.000" long.
- ❏ 6. Turn the portion that will become the taper to 1.250" diameter × 1.000" length.
- ❏ 7. Turn the large end of the workpiece to 1.750" diameter.
- ❏ 8. Set up the compound and cut the 30° angle.
- ❏ 9. Refer to the *Machining Fundamentals* textbook and *Machinery's Handbook* to figure the degree of taper for the tapered section of the tool holder.
- ❏ 10. Set the compound or the taper attachment for the correct angle and cut the taper.
- ❏ 11. Set up the lathe with a 4-jaw chuck and using a dial indicator, center the tool holder, and drill and ream the shank hole to the desired diameter. This dimension can be tailored to the size end mill shank for which it will be used.
- ❏ 12. Reverse the tool holder in the 4-jaw chuck, indicate the shank, then drill 25/64" × 2.5" deep, then tap 7/16-20.
- ❏ 13. Place the tool holder in the vise on the vertical mill. Be certain to check the trim of the mill head and the alignment of the vise before making any cuts.
- ❏ 14. Mount the correct size end mill, touch off and set vertical adjustment, and cut the keyway to the specified length and depth.
- ❏ 15. Locate the center of the set screw hole and drill and tap as shown.
- ❏ 16. Break all sharp edges and install set screw.

## Notes

_____

_____

_____

---

**Performance Evaluation—Instructor's Use Only**

Project completed on time?   ❏ Yes   ❏ No

   If No, which steps were not completed _____

Overall performance on project:

   ❏ Excellent   ❏ Good   ❏ Satisfactory   ❏ Unsatisfactory   ❏ Poor

Comments: _____

_____

Instructor's Signature _____

# Project 4.10 R-8 Tool Holder

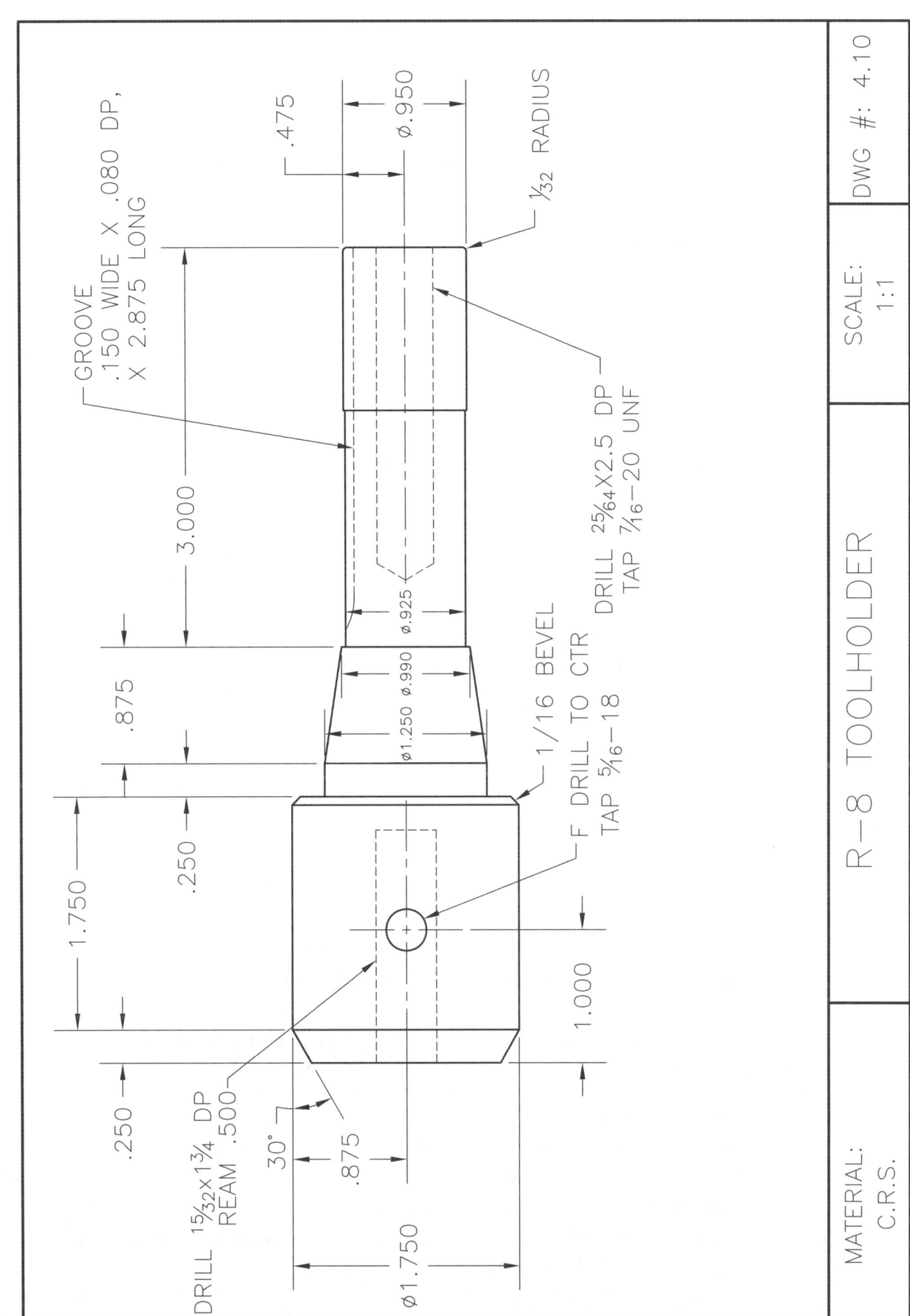

# Project 4.11 Machinist Screw Jack

Name _____ Date _____

Instructor _____ Period _____

## Order of Operations

### Base

❏ 1. Measure and cut stock plus facing allowance.

❏ 2. Calculate correct speed and feed for the type and size material being used and the type and size lathe tool bit.

❏ 3. Face to the specified overall length.

❏ 4. Center drill one end.

❏ 5. Calculate correct speed required for drilling the type material being used and the size drill bit specified.

❏ 6. Drill 1″ diameter through. It is recommended that the piece be step drilled with drill bits of progressively larger diameters until the required diameter is achieved.

❏ 7. On one end, make the interior cut to the depth and diameter specified on the print, then chamfer the edge.

❏ 8. Turn the workpiece around, adjust the lathe for straight turning, and cut this end to the length and diameter required.

❏ 9. Set up the lathe for knurling and knurl the section indicated with a medium knurl.

❏ 10. Make the under cut.

❏ 11. Chamfer remaining edges.

❏ 12. Set up the lathe for chasing interior threads and cut the 1-14 thread to the required depth.

### Telescoping Screw

❏ 1. Cut a piece of stock approximately 4 1/2″ long. If 1 5/16″ diameter stock is not on hand, use a slightly larger diameter and turn to the specified diameter.

❏ 2. Face and center drill both ends.

❏ 3. Knurl a section approximately 2″ long in the middle of the piece.

❏ 4. Turn the section to be threaded to 1″ and cut or file the chamfer on the end.

❏ 5. Cut the relief groove as shown.

❏ 6. Set up the lathe for chasing threads and cut the 1-14 thread. Use the base to check for fit. Before attempting to screw the base onto the telescope screw, be certain that both pieces are clean and free of chips. Each piece should be lightly oiled to prevent seizing. Repeat chasing cuts until a Class 1 fit is achieved.

❏ 7. With the appropriate tap drill, drill the workpiece to a full diameter depth of 1 7/8″.

❏ 8. Countersink to the major diameter of the thread and tap 5/8-18.

❏ 9. Cut or part the workpiece at least 1 3/4″ from the end.

❏ 10. Turn the piece around and face to print specifications. Be certain to protect the 1-14 threads when chucking for the facing operation.

*Machining Projects*      **Project 4.11**      Machinist Screw Jack

## Screw

- ❏ 1. Cut a piece of 1 1/8" diameter stock approximately 3" long.
- ❏ 2. Face and center drill both ends.
- ❏ 3. On one end, knurl a section about 1" long.
- ❏ 4. Turn the piece around and chuck it by 1/4" of the end, support to other end with a live center, and turn the portion to be threaded to the dimensions indicated on the print. Cut or file the chamfer after the correct length and diameter have been attained.
- ❏ 5. Cut the relief groove.
- ❏ 6. Set up the lathe for chasing threads and cut the 18 threads per inch to appropriate depth. Use the telescope screw to check for fit. Before attempting to try these two parts, be certain the threaded portions of both are free of chips and that they are lightly lubricated to prevent seizing. Continue chasing cuts until a Class 1 fit is achieved.
- ❏ 7. Turn the workpiece around and face to print specifications. Do not forget to protect the threads when chucking on this end.
- ❏ 8. Turn the section that will become the ball to the specified length and diameter.
- ❏ 9. Make 3/8" under cut.
- ❏ 10. Cut or file the ball to the size indicated.
- ❏ 11. Chamfer both edges of the knurled section of the screw.
- ❏ 12. Set up the vertical mill or drill press, locate the center of the hole and drill and countersink as shown.

## Cap

- ❏ 1. Measure and cut stock plus facing allowance for the cap.
- ❏ 2. Face to the specified length and center drill one end.
- ❏ 3. Drill as shown on the print.
- ❏ 4. Cut the bevel.
- ❏ 5. Reverse the piece in the chuck and cut or file the chamfer.
- ❏ 6. Again reverse the cap and chuck with about 1/2" of its length extending past the face of the chuck jaws.
- ❏ 7. Place the screw in a Jacobs chuck mounted in the tailstock spindle and move forward until the ball portion of the screw is fully in the 9/16" hole in the cap.
- ❏ 8. Using a lathe tool bit that has been ground to a smooth radius, turn on the lathe, and advance to rounded side of the tool bit until it contacts the edge of the cap.
- ❏ 9. Apply pressure until the thin edge of the bevel has been rolled in enough to capture the ball.
- ❏ 10. Clean all parts of the screw jack and assemble.

## Notes

_____

_____

_____

---

**Performance Evaluation—Instructor's Use Only**

Project completed on time?    ❏ Yes    ❏ No

    If No, which steps were not completed _____

Overall performance on project:

    ❏ Excellent    ❏ Good    ❏ Satisfactory    ❏ Unsatisfactory    ❏ Poor

Comments: _____

_____

Instructor's Signature_____

**Project 4.11** Machinist Screw Jack *Machining Projects*

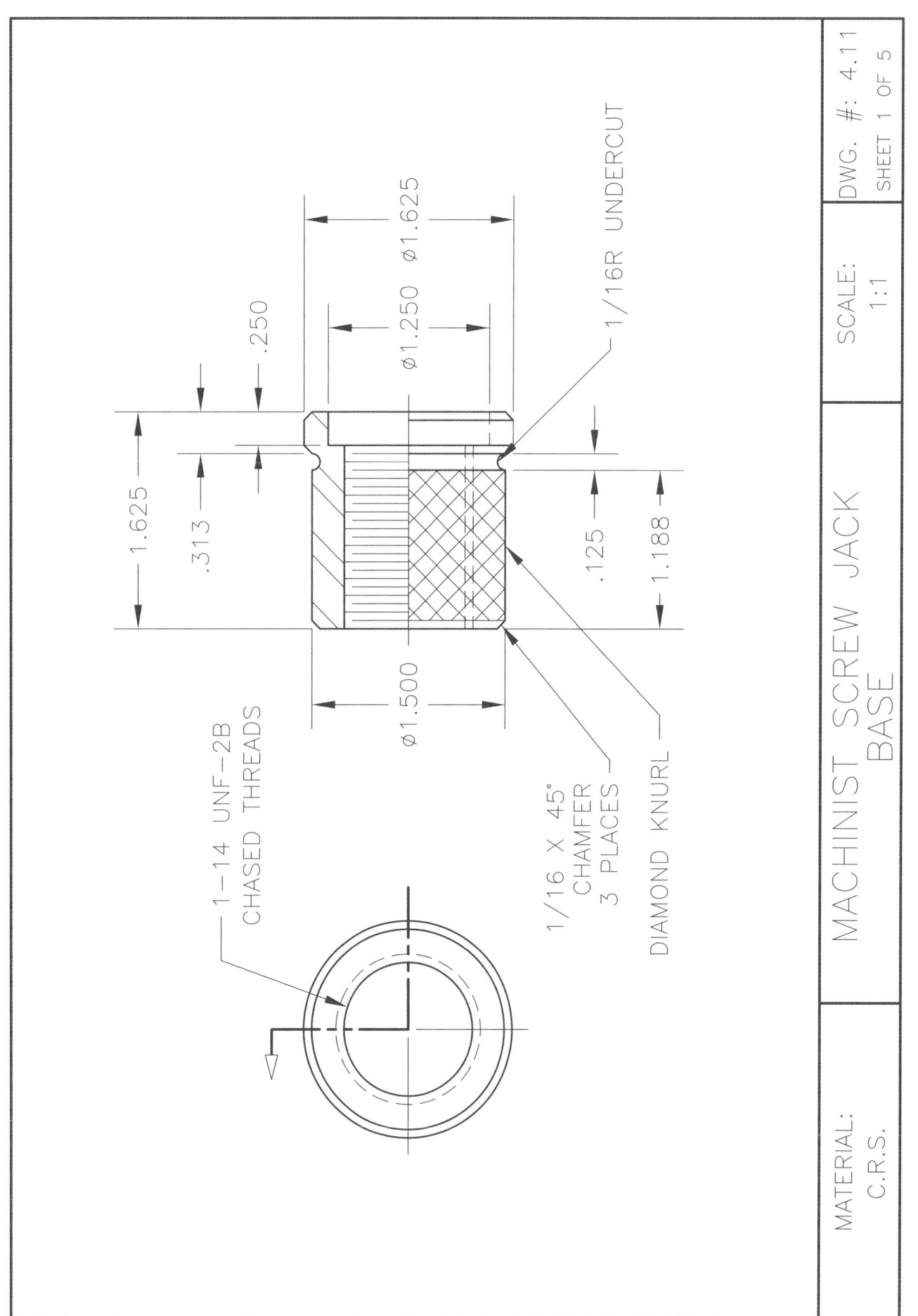

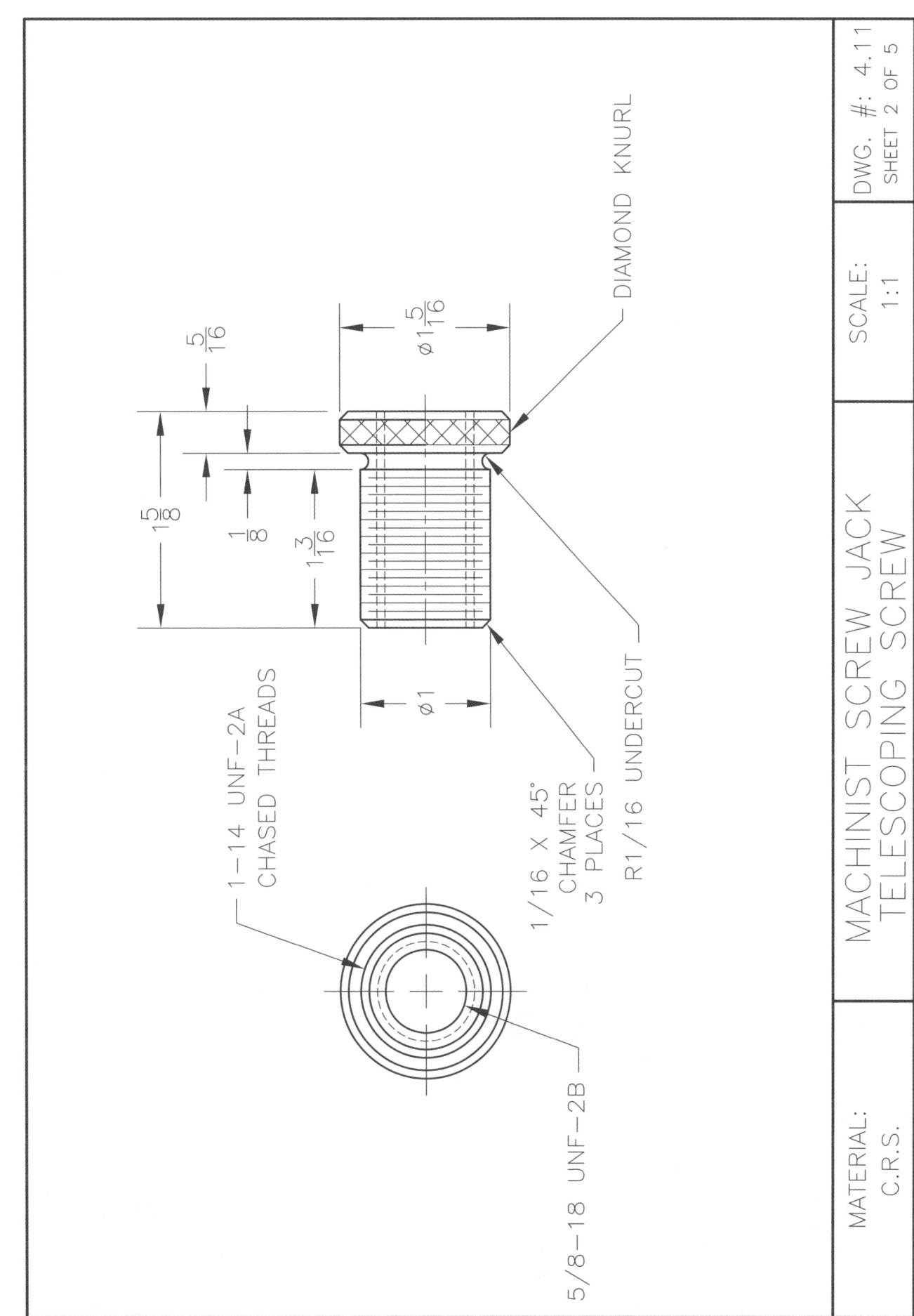

## Project 4.11 Machinist Screw Jack

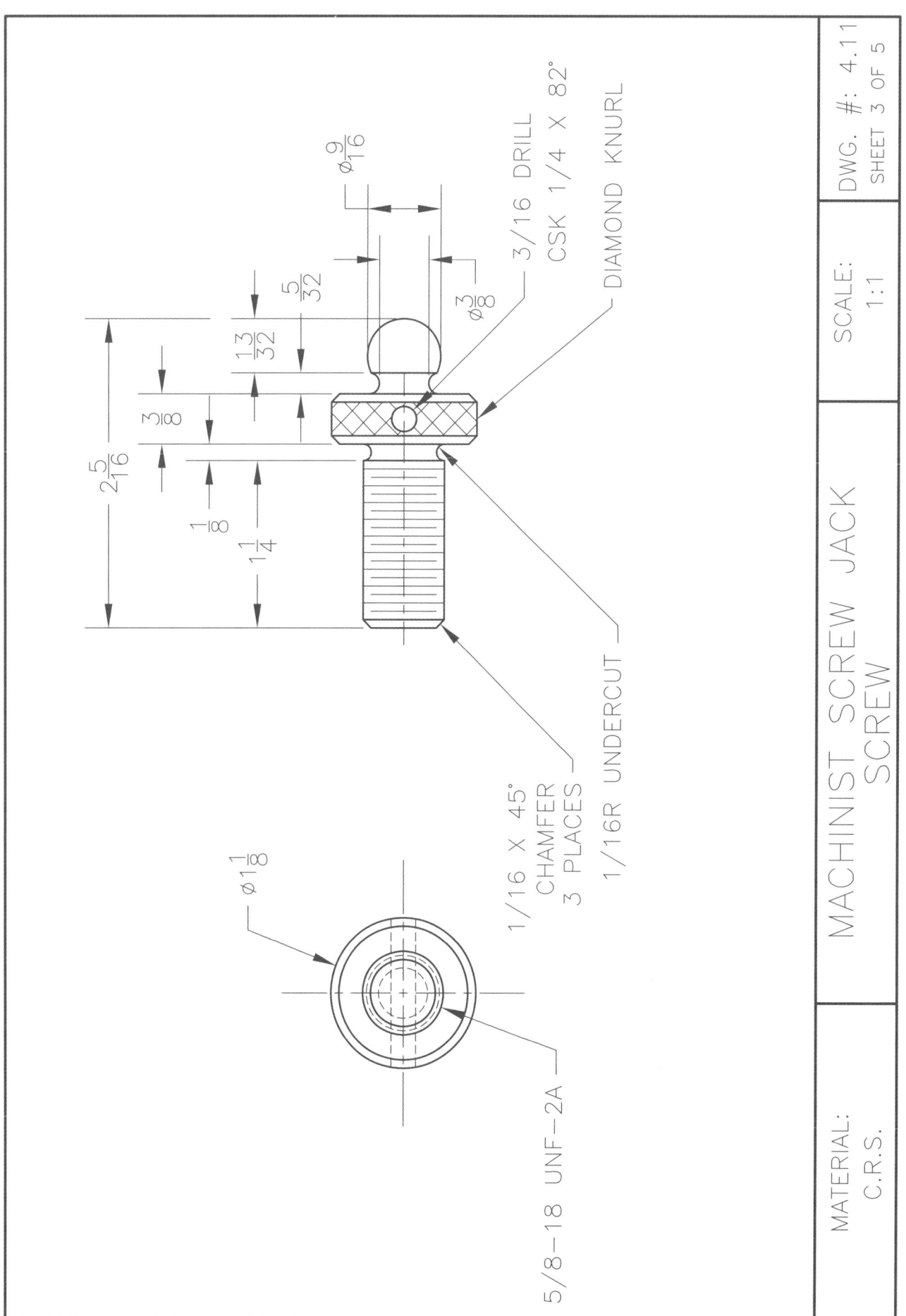

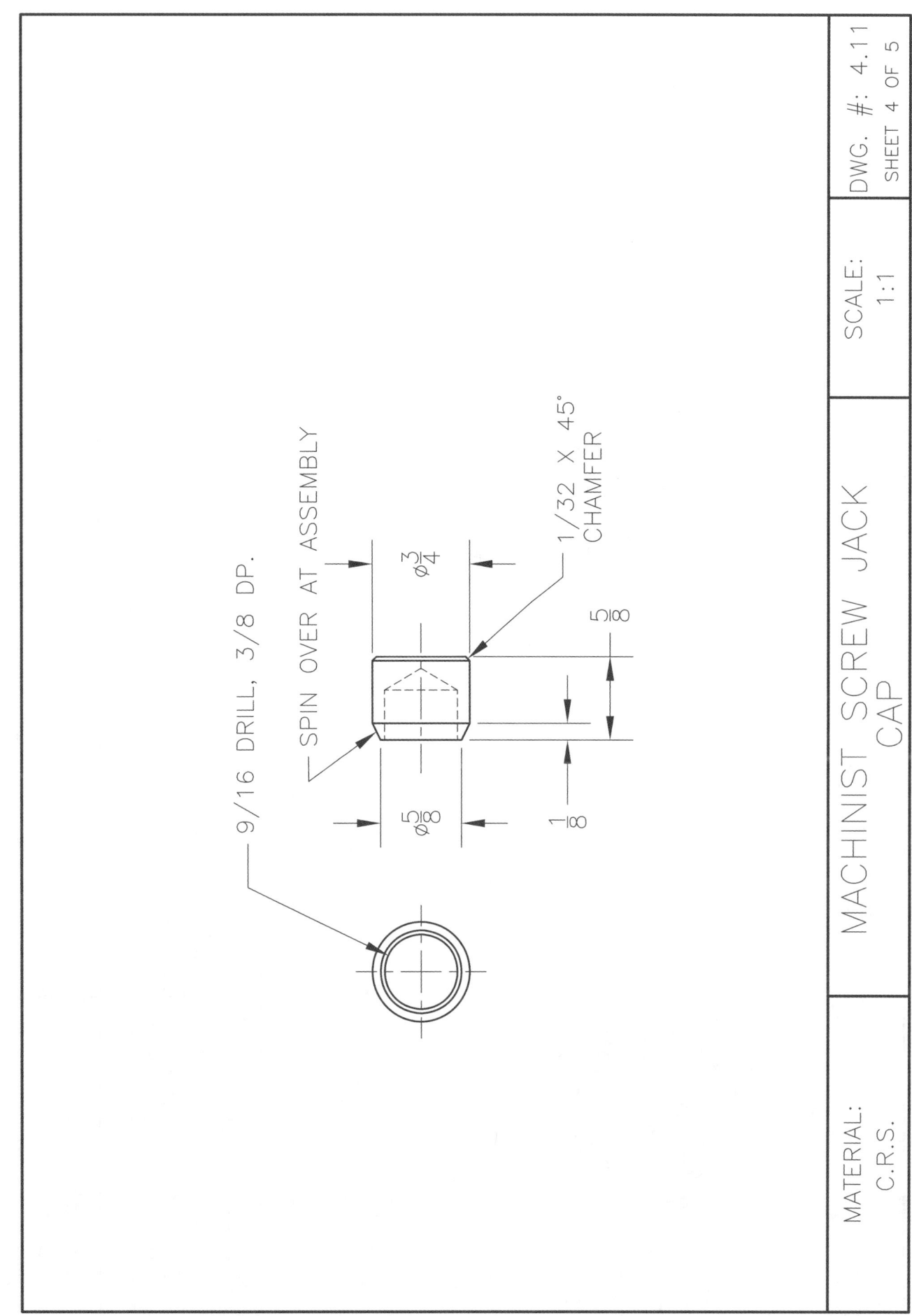

**Project 4.11**   Machinist Screw Jack

*Machining Projects*

Ø1.625

3.00 MIN.
5.3125 MAX.

Ø.750

| PART #. | DWG. #. | DESCRIPTION | MATERIAL: | QTY. REQD. |
|---|---|---|---|---|
| 4 | 4.11, SH4 | CAP | STEEL | 1 |
| 3 | 4.11, SH3 | SCREW | STEEL | 1 |
| 2 | 4.11, SH2 | TELESCOPE SCREW | STEEL | 1 |
| 1 | 4.11, SH1 | BASE | STEEL | 1 |

PARTS LIST

| MATERIAL: AS NOTED. | MACHINIST SCREW JACK ASSEMBLY | SCALE: 1:1 | DWG. #: 4.11 SHEET 5 OF 5 |
|---|---|---|---|

138

# Project 4.12
## Meat Tenderizing Hammer

Name _____ Date _____

Instructor _____ Period _____

## Order of Operations

**Tenderizer Head**

☐ 1. Cut appropriate stock plus 3/16" facing allowance.

☐ 2. Calculate the correct cutting speed and feed for the type and diameter end mill to be used.

☐ 3. Face the workpiece to the specified length.

☐ 4. Using an edge finder, locate, drill, counterbore, countersink, and tap the 1/2-13 hole.

☐ 5. Mount the workpiece in the vise of the vertical or horizontal mill and cut the teeth as indicated on the print.

    **Note** Step 5 can be easily accomplished using Project 4.8 Mill Cutter Arbor.

☐ 6. Rotate the workpiece 90° in the vise and make another series of cuts to create the pyramid-shaped teeth.

☐ 7. Perform the same operations described above on the other end of the workpiece.

**Handle**

☐ 1. Cut the appropriate aluminum stock for the handle. Remember to include about 3/16" facing allowance when cutting the piece.

☐ 2. Calculate the correct cutting speed and feed for the type of lathe tool bit to be used.

☐ 3. Face the workpiece to the correct length.

☐ 4. Straight turn all outside diameters as indicated.

☐ 5. Drill the back end of the handle 19/32", then ream 5/8".

☐ 6. Thread the front end of the hammer handle 1/2-13UNC as shown.

☐ 7. Calculate the tailstock setover, taper per inch, or the degree of taper according to the method to be used for turning the tapers.

☐ 8. Set up the lathe for the taper turning method to be used.

☐ 9. Cut the tapers.

☐ 10. Set up the dividing head on the table of a vertical mill. Use a round bar and dial indicator to ensure alignment.

☐ 11. Calculate correct speed and feed for the type and size end mill to be used.

☐ 12. Cut the eight equally spaced flutes as shown on the drawing.

**Project 4.12**  Meat Tenderizing Hammer

## Plug

- [ ] 1. Cut a short (approx. 2″) piece of 3/4″ diameter piece of aluminum and face one end.
- [ ] 2. Turn a section on the end 1/4″ long × .002″ larger than the *actual* hole diameter in the end of the handle and file a slight bevel on the end.
- [ ] 3. Part off approximately 3/4″ of the piece.
- [ ] 4. Turn the piece around, chuck it by the turned down section, and face to correct overall length.
- [ ] 5. Cut or file the chamfer as indicated on the print.
- [ ] 6. Assemble head and handle, then press in the plug using an arbor press.

## Notes

---

**Performance Evaluation—Instructor's Use Only**

Project completed on time?  ❏ Yes  ❏ No
  If No, which steps were not completed _____
Overall performance on project:
  ❏ Excellent  ❏ Good  ❏ Satisfactory  ❏ Unsatisfactory  ❏ Poor
Comments: _____
_____
Instructor's Signature_____

*Machining Projects* — **Project 4.12** — Meat Tenderizing Hammer

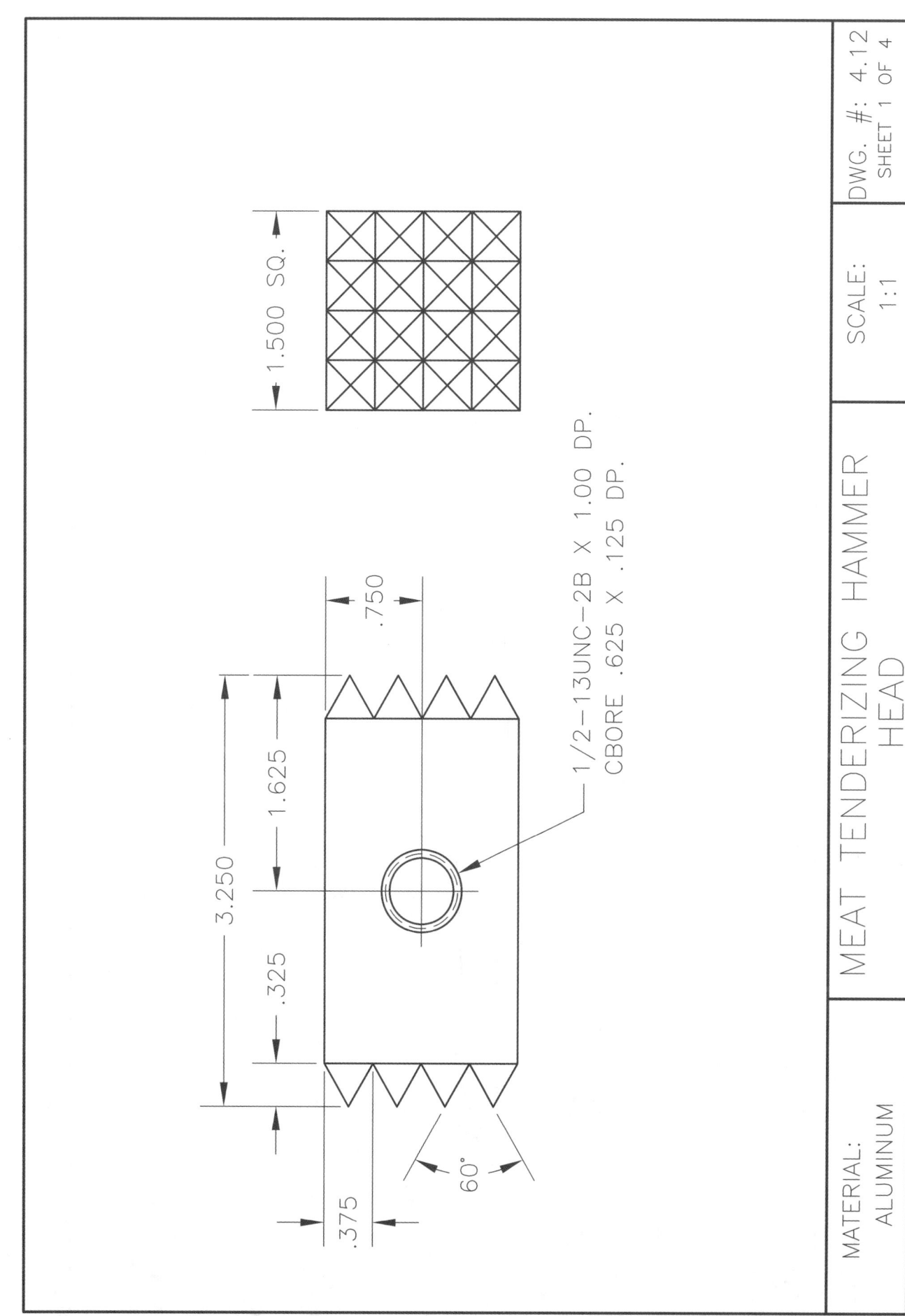

**Project 4.12** Meat Tenderizing Hammer

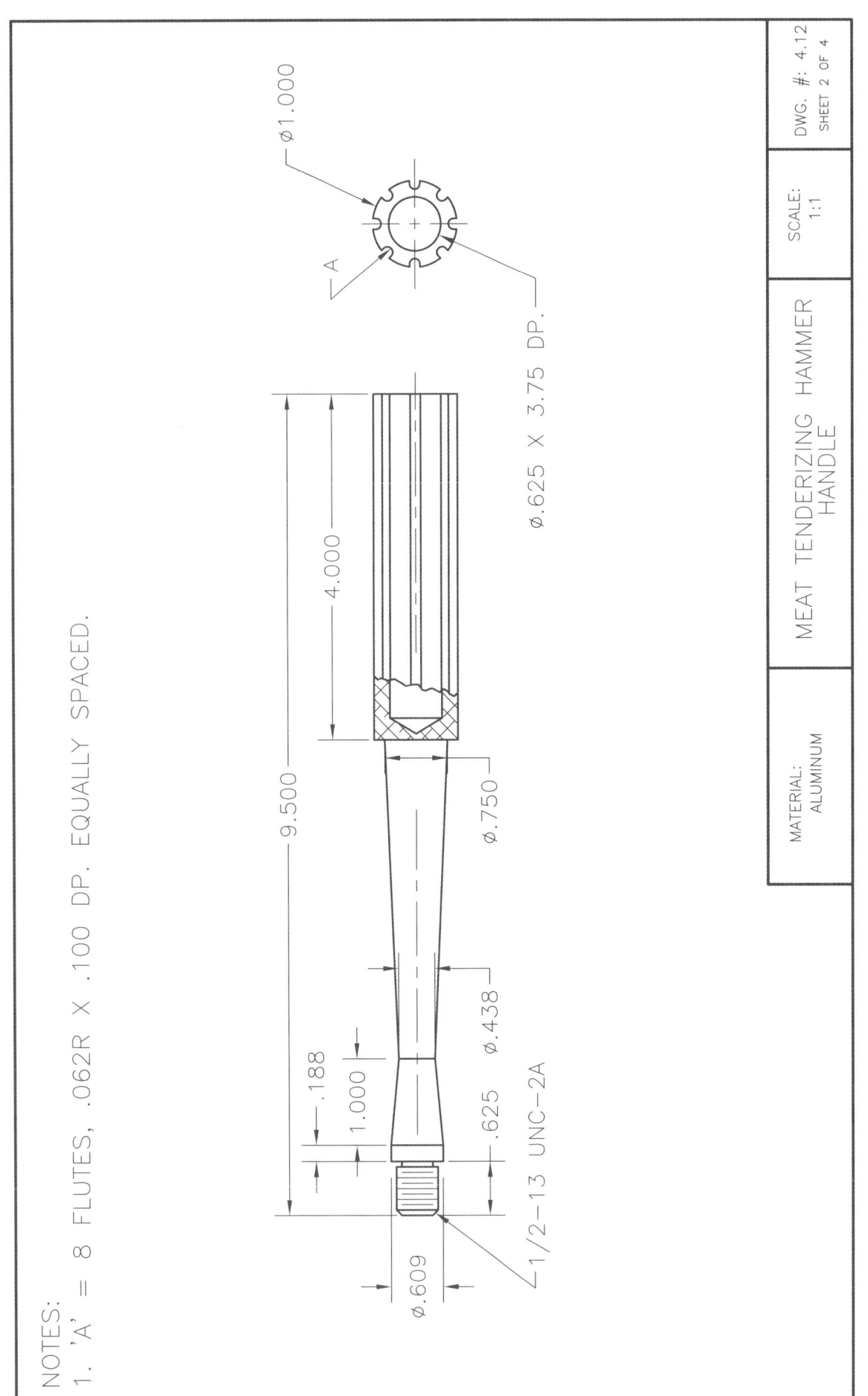

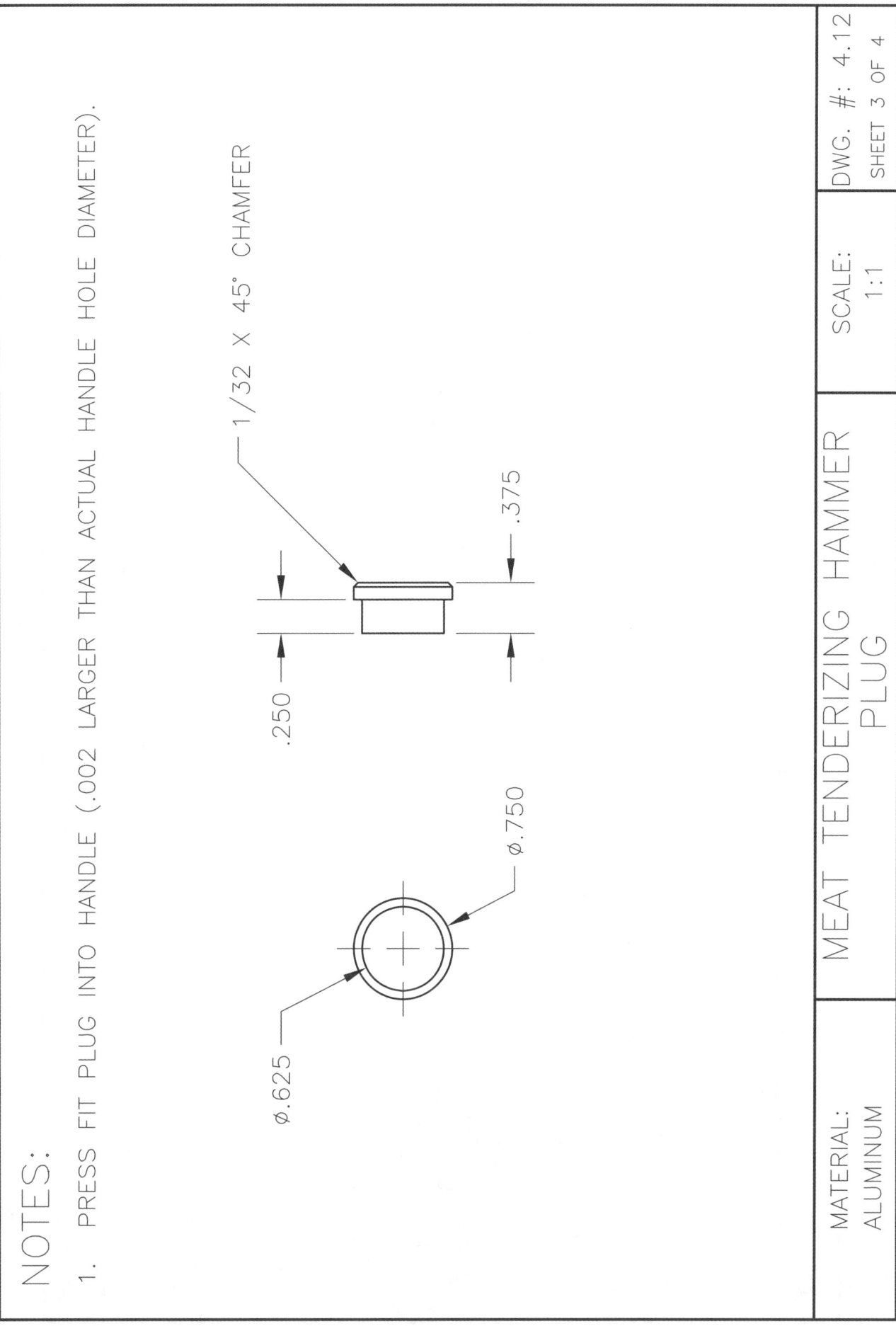

# Project 4.12  Meat Tenderizing Hammer

*Machining Projects*

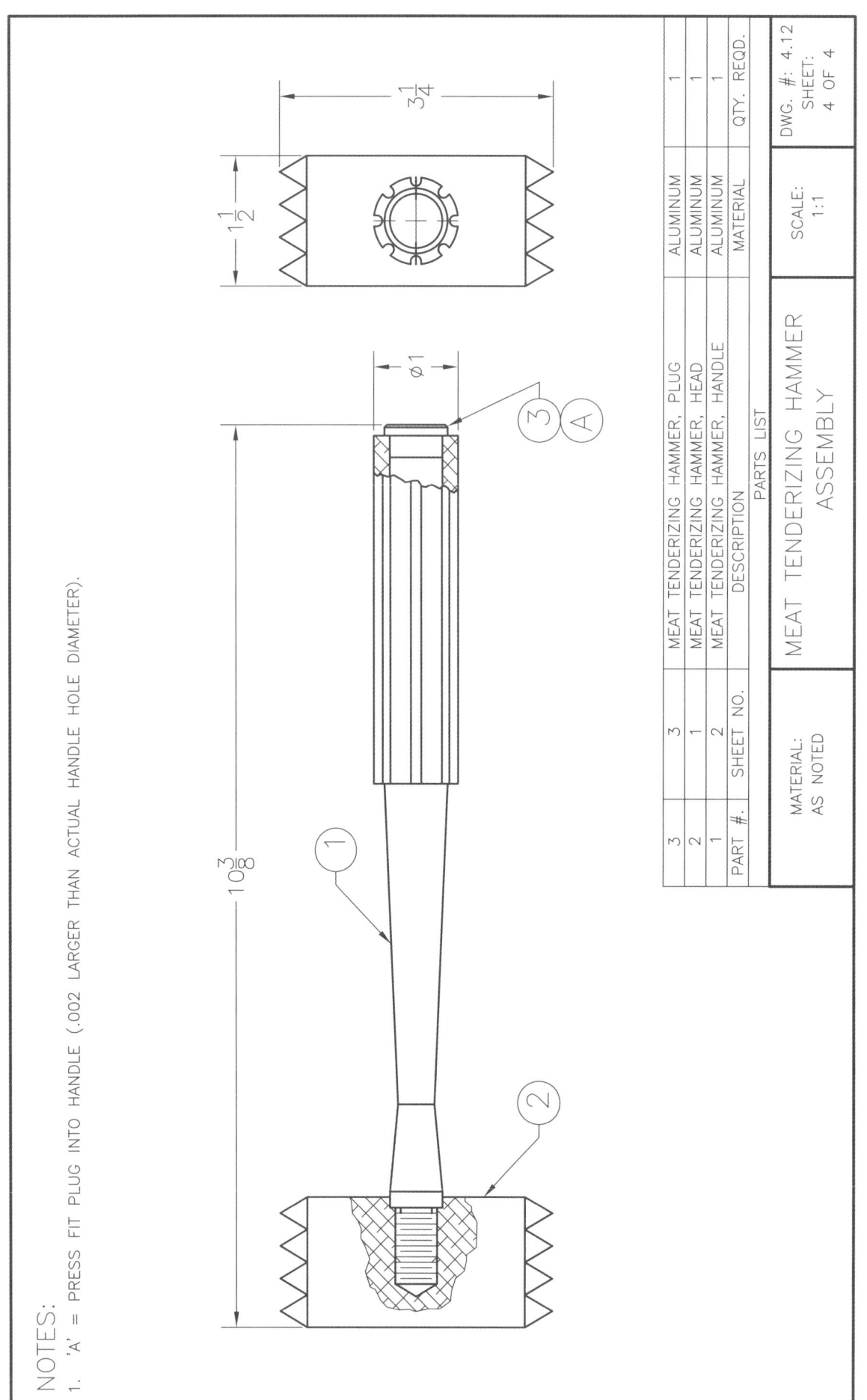

NOTES:
1. 'A' = PRESS FIT PLUG INTO HANDLE (.002 LARGER THAN ACTUAL HANDLE HOLE DIAMETER).

| PART #. | SHEET NO. | DESCRIPTION | MATERIAL | QTY. REQD. |
|---|---|---|---|---|
| 3 | 3 | MEAT TENDERIZING HAMMER, PLUG | ALUMINUM | 1 |
| 2 | 1 | MEAT TENDERIZING HAMMER, HEAD | ALUMINUM | 1 |
| 1 | 2 | MEAT TENDERIZING HAMMER, HANDLE | ALUMINUM | 1 |

PARTS LIST

MEAT TENDERIZING HAMMER ASSEMBLY

MATERIAL: AS NOTED

SCALE: 1:1

DWG. #: 4.12
SHEET: 4 OF 4

# Project 4.13
# Paper Punch

Name _____  Date _____

Instructor _____  Period _____

## Order of Operations

### Body
❏ 1. Measure and cut stock plus squaring allowance.
❏ 2. Check trim and alignment of the vertical mill head and vise with the dial indicator.
❏ 3. Install an appropriate end mill.
❏ 4. Calculate the correct cutting speed and feed for the type and size end mill to be used and the type of material.
❏ 5. Using correct set up and technique, mill the body to print specifications.
❏ 6. Use an edge finder to locate the position of the 3/4" diameter hole.
❏ 7. Mount the drill chuck, center drill, and drill the hole.
❏ 8. Reposition the workpiece and drill and tap the 1/4-20 holes.
❏ 9. Remove the drill chuck, mount an end mill, and cut the 1/8" step as shown.
❏ 10. Reposition the piece in the vise and mill the 1/2" deep × 1 1/8" wide area.

### Base
❏ 1. Measure and cut stock plus facing allowance.
❏ 2. Mill the base to 2" square.
❏ 3. Using an edge finder, locate, drill, and countersink the two holes shown.

### Plunger
❏ 1. Measure and cut stock plus facing allowance.
❏ 2. Face plunger to the length specified.
❏ 3. Knurl approximately 3/8" at one end of the piece.
❏ 4. Locate and drill the 1/16" diameter hole as indicated.
❏ 5. Mill or file the 5° angle on the opposite end. The angle should have a clean, sharp edge.

### Head
❏ 1. Face one end of a piece of 1 1/2" diameter stock of the kind specified.
❏ 2. Turn the end of the piece to 1/2" diameter by 3/8" long.
❏ 3. Drill and ream the 1/4" diameter hole to the depth shown.
❏ 4. Part or cut the piece off, turn it around in the chuck, and face to the required length.
❏ 5. Using a file, round this end of the head to the shape shown on the print.

**Project 4.13**  Paper Punch  *Machining Projects*

## Spring
- [ ] 1. Wrap .045″ diameter music wire tightly around a 1/4″ diameter mandrel. The plunger can be used for this operation.
- [ ] 2. After wrapping, trim the resulting spring to the specified length.

## Washer
- [ ] 1. Measure, cut, and face piece of 1/2″ diameter stock.
- [ ] 2. Center drill and drill one end of the piece with a 7/32″ drill and then ream to 1/4″ diameter.
- [ ] 3. Cut or part off the washer.
- [ ] 4. Cut or file a slight chamfer on both sides.

## Assembly Operations
- [ ] 1. Using an arbor press, press the knurled end of the plunger into the head.
- [ ] 2. Secure the body to the base with two 1/4-20 flat head machine screws.
- [ ] 3. Position the assembled parts in the vise of a vertical mill.
- [ ] 4. With an edge finder, locate, center drill, drill, and ream the 1/4″ diameter hole.
- [ ] 5. Assemble remaining parts.

## Notes

---

**Performance Evaluation—Instructor's Use Only**

Project completed on time?   ❏ Yes   ❏ No
    If No, which steps were not completed _____
Overall performance on project:
    ❏ Excellent   ❏ Good   ❏ Satisfactory   ❏ Unsatisfactory   ❏ Poor
Comments: _____

Instructor's Signature _____

*Machining Projects* **Project 4.13** Paper Punch

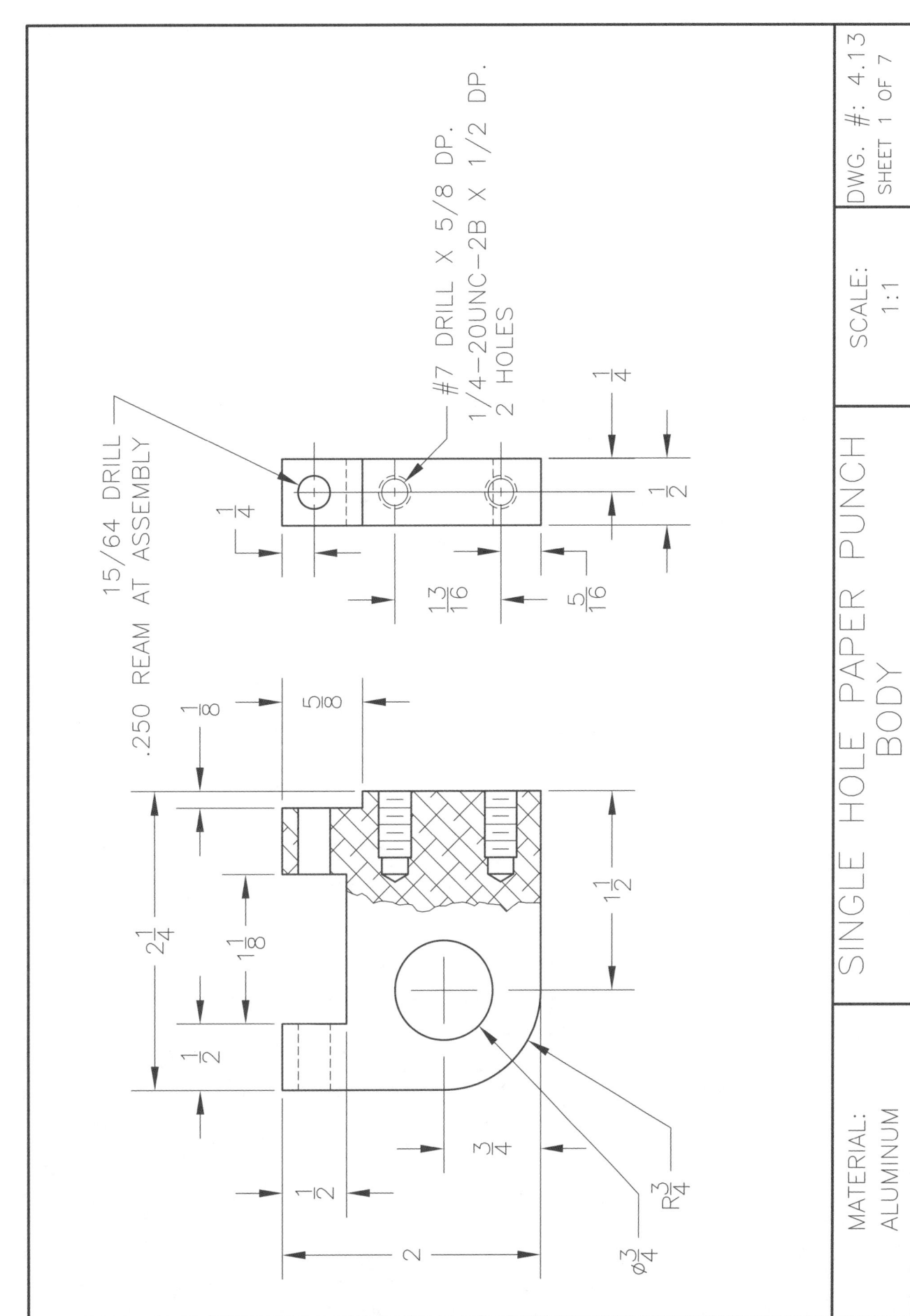

**Project 4.13** Paper Punch

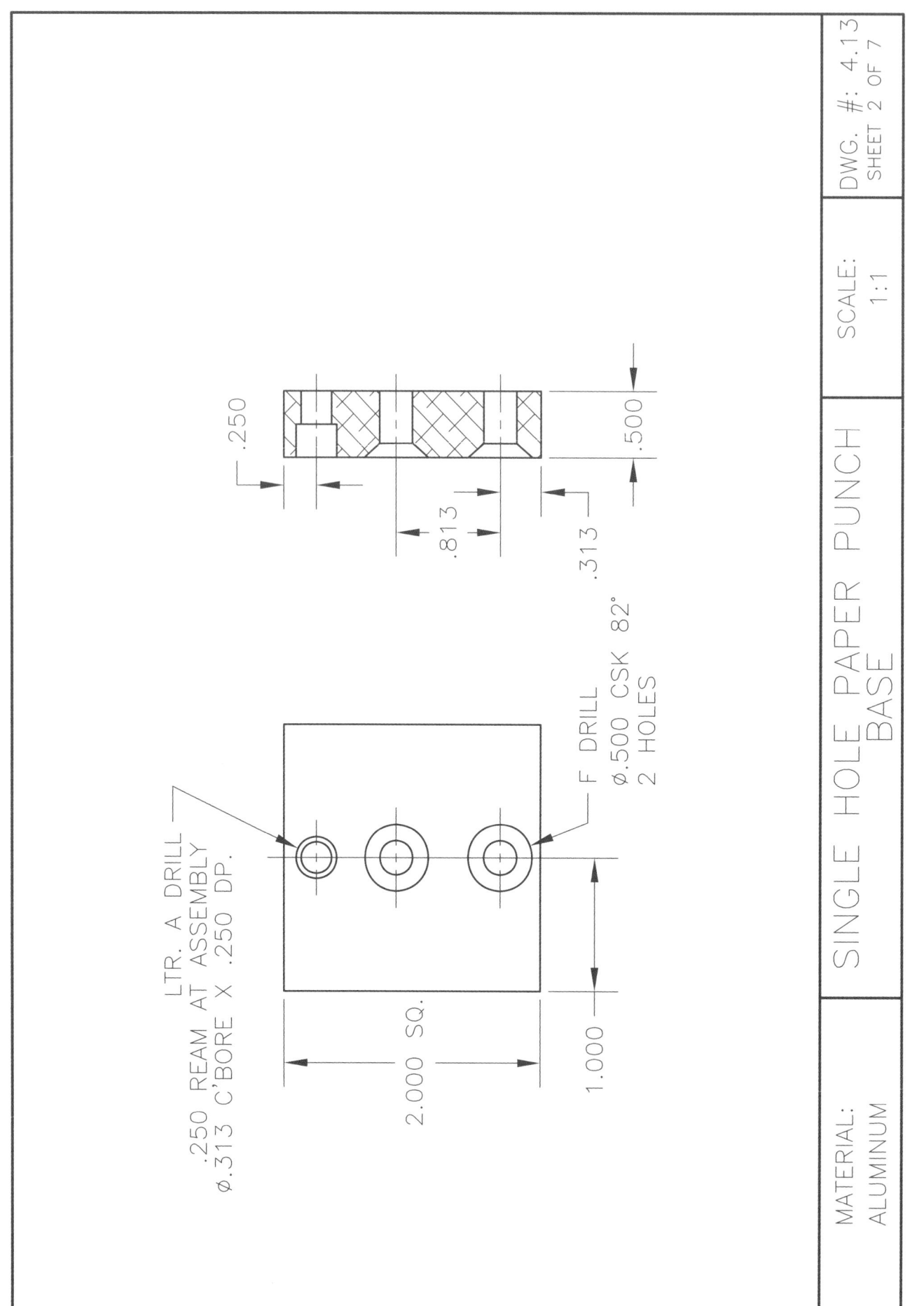

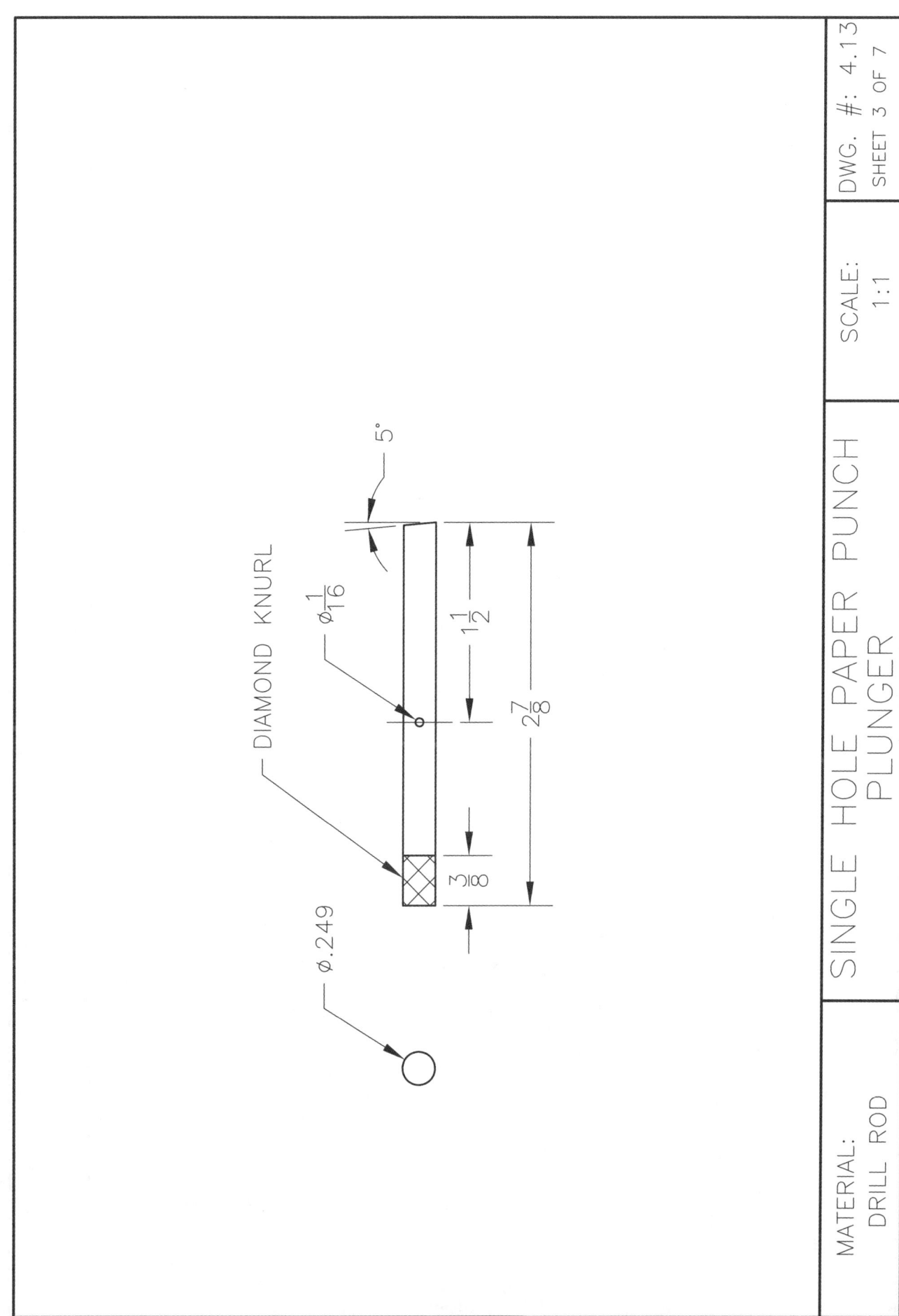

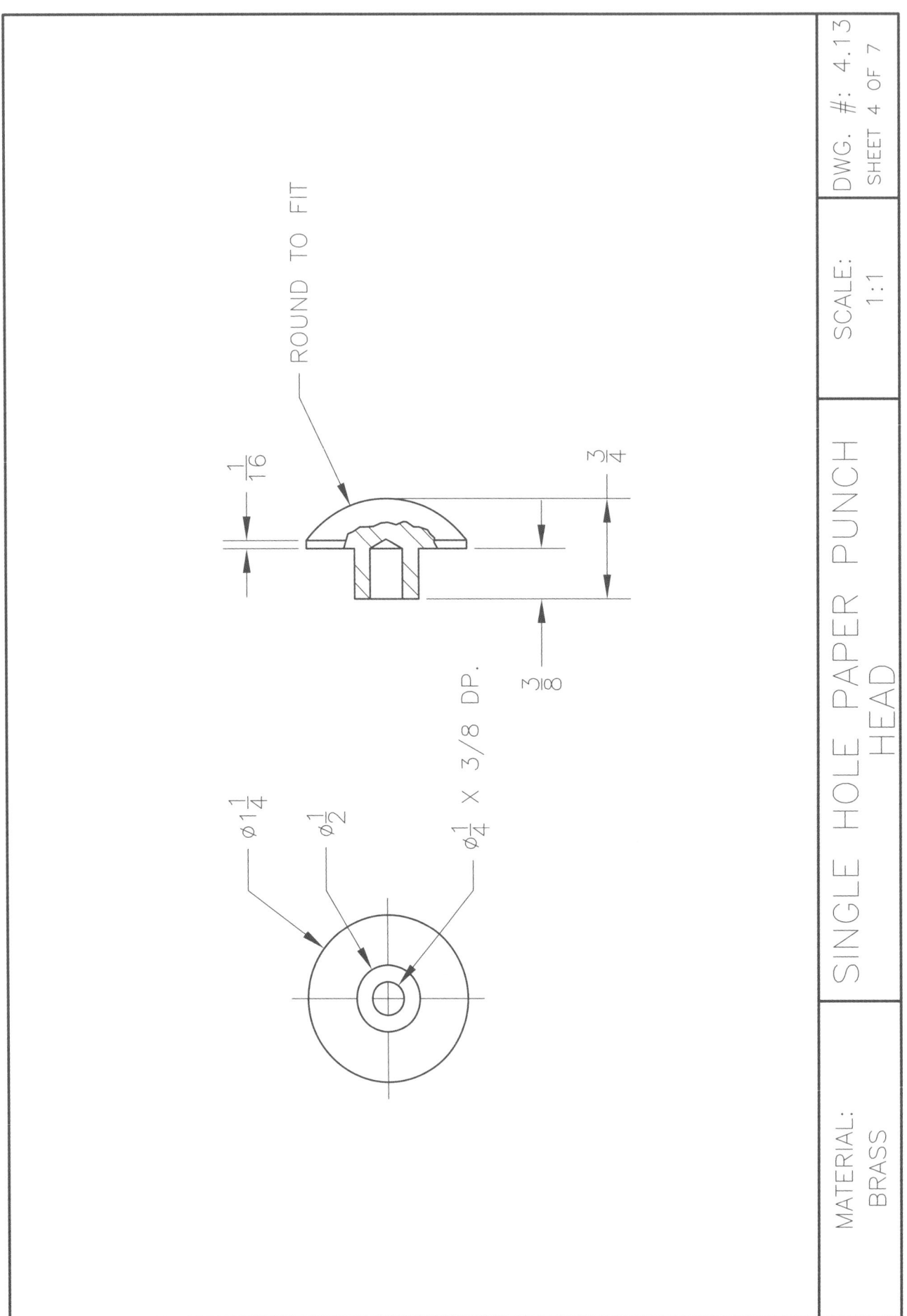

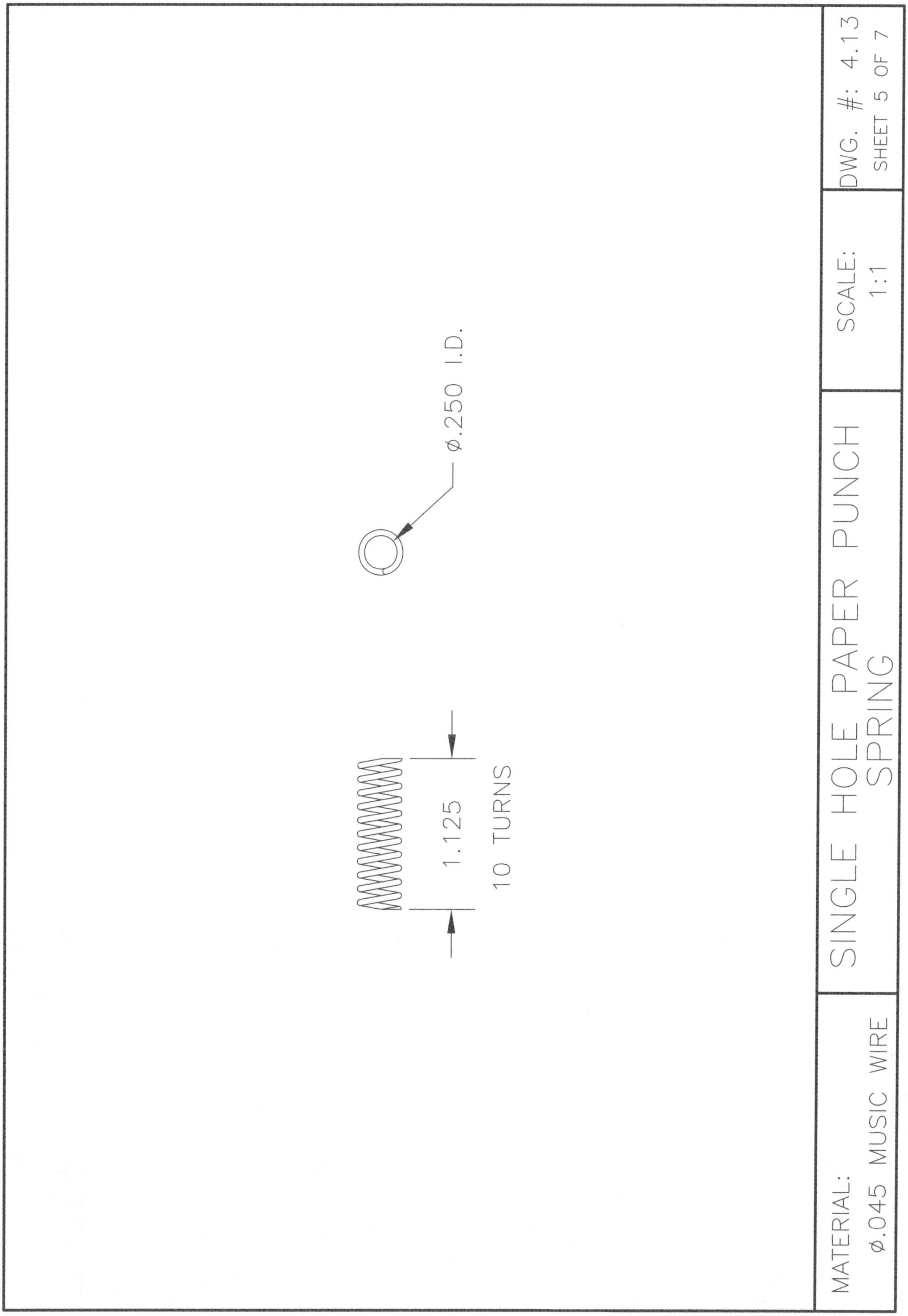

Machining Projects    **Project 4.13**    Paper Punch

⌀.250 I.D.

1.125
10 TURNS

DWG. #: 4.13
SHEET 5 OF 7

SCALE: 1:1

SINGLE HOLE PAPER PUNCH SPRING

MATERIAL:
⌀.045 MUSIC WIRE

**Project 4.13** Paper Punch  *Machining Projects*

$\varnothing\frac{1}{4}$ REAM

$\varnothing\frac{1}{2}$

$\frac{1}{8}$

1/32 × 45° CHAMFER BOTH SIDES

| | |
|---|---|
| MATERIAL: BRASS | SINGLE HOLE PAPER PUNCH WASHER |

SCALE: 1:1

DWG. #: 4.13
SHEET 6 OF 7

# Machining Projects

## Project 4.13    Paper Punch

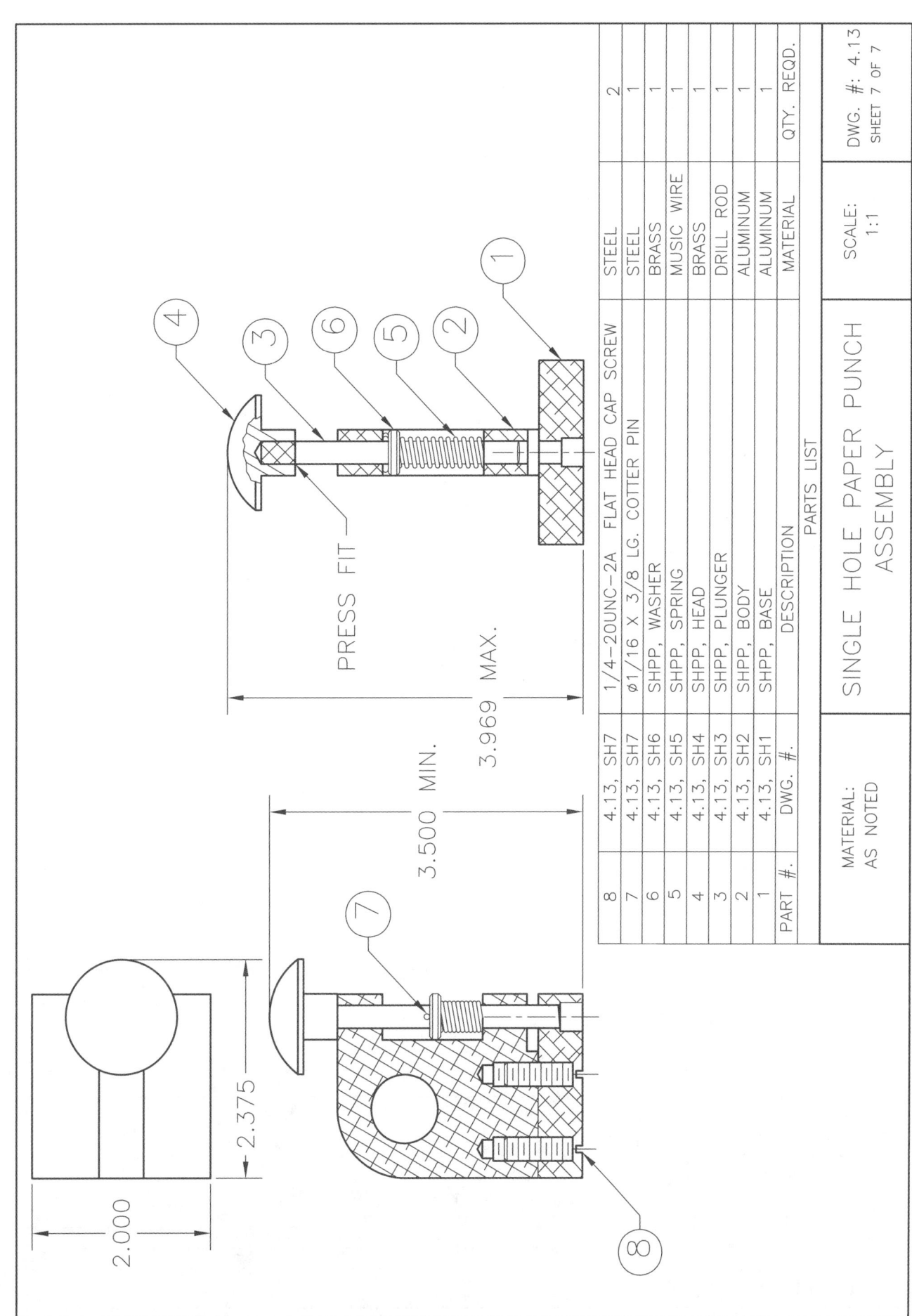

**Project 4.14** Tap Wrench

# Project 4.14
# Tap Wrench

Name _____ Date _____

Instructor _____ Period _____

## Order of Operations

### Handles
- ❏ 1. Measure and cut stock plus facing allowance. Two pieces are required.
- ❏ 2. Face pieces to the length specified on the print.
- ❏ 3. Center drill one end of each piece and then knurl for 2 1/2".
- ❏ 4. Turn down the knurled ends and thread 5/16-18 for 1".

### V-Blocks
- ❏ 1. Measure and cut stock plus facing allowance. Two pieces are required.
- ❏ 2. Face both pieces to the length indicated on the print.
- ❏ 3. Locate and drill the blocks as shown.
- ❏ 4. Tap the appropriate hole on each block 5/16-18.
- ❏ 5. Cut the 90° V-notch to the width specified.
- ❏ 6. Cut or file the 1/16" by 45° bevels on each block.
- ❏ 7. Polish to an acceptable finish and assemble.

## Notes

_____
_____
_____
_____
_____
_____
_____
_____
_____

---

**Performance Evaluation—Instructor's Use Only**
Project completed on time?    ❏ Yes    ❏ No
  If No, which steps were not completed _____
Overall performance on project:
  ❏ Excellent    ❏ Good    ❏ Satisfactory    ❏ Unsatisfactory    ❏ Poor
Comments: _____
_____

Instructor's Signature _____

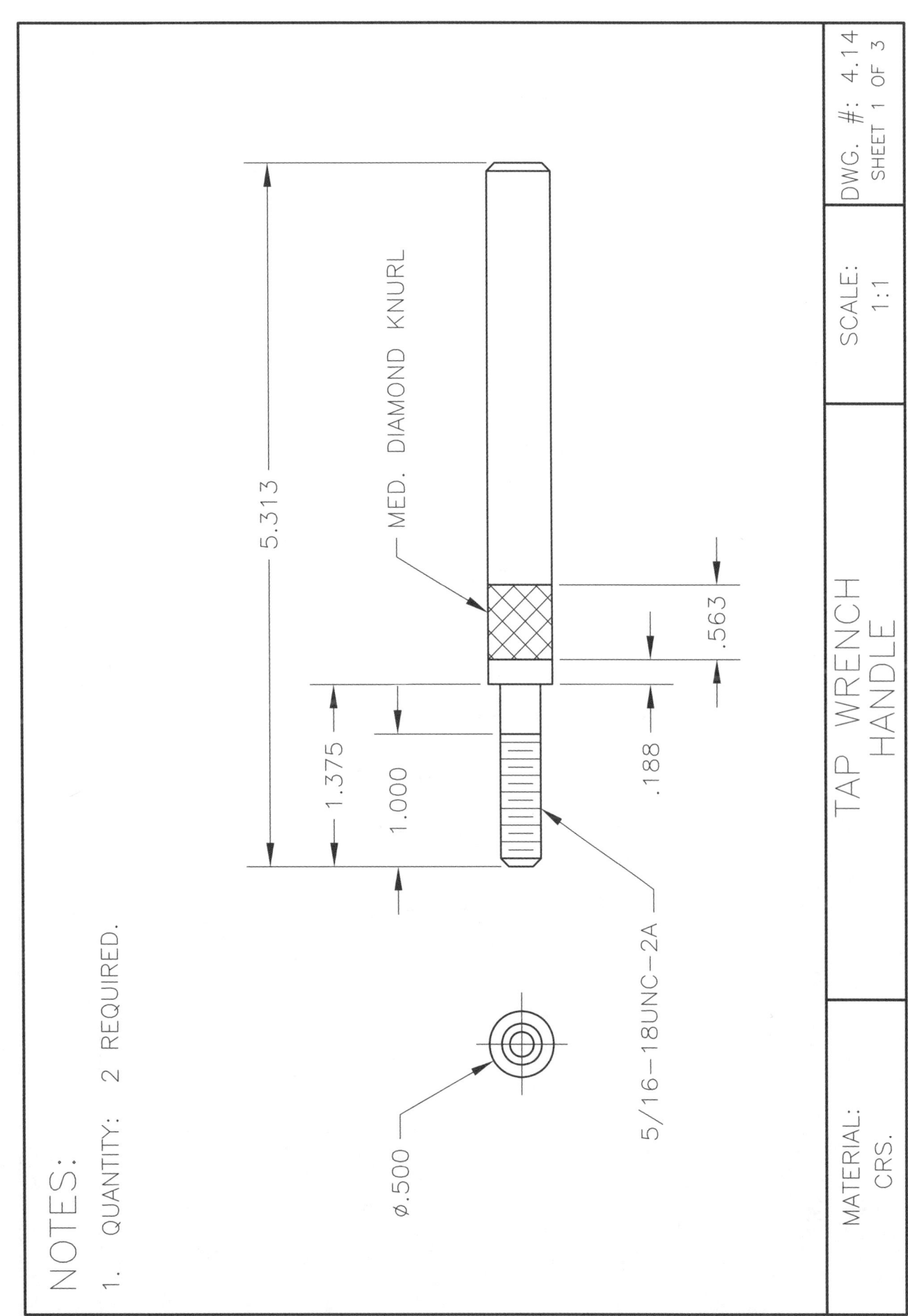

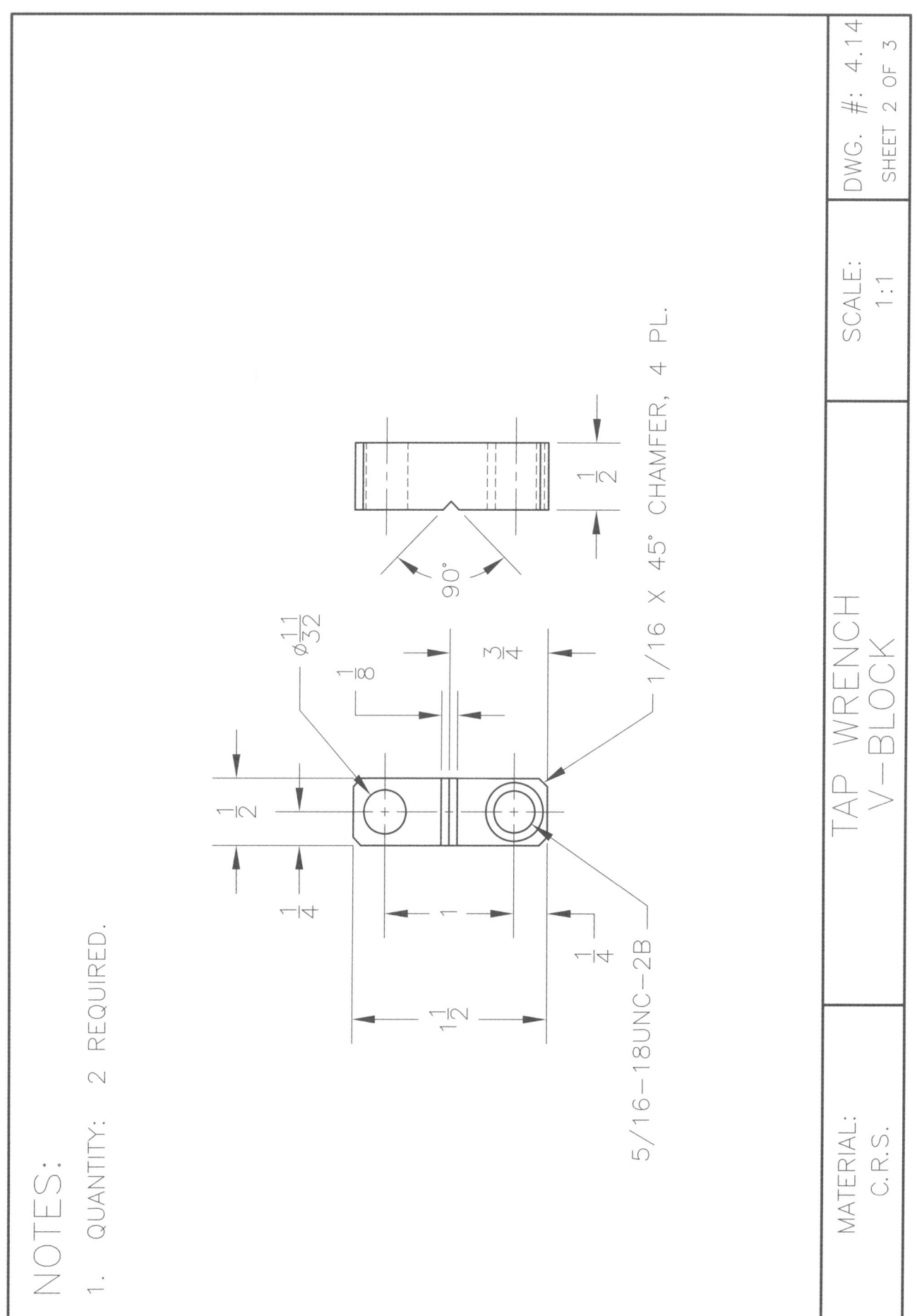

# Project 4.14 Tap Wrench

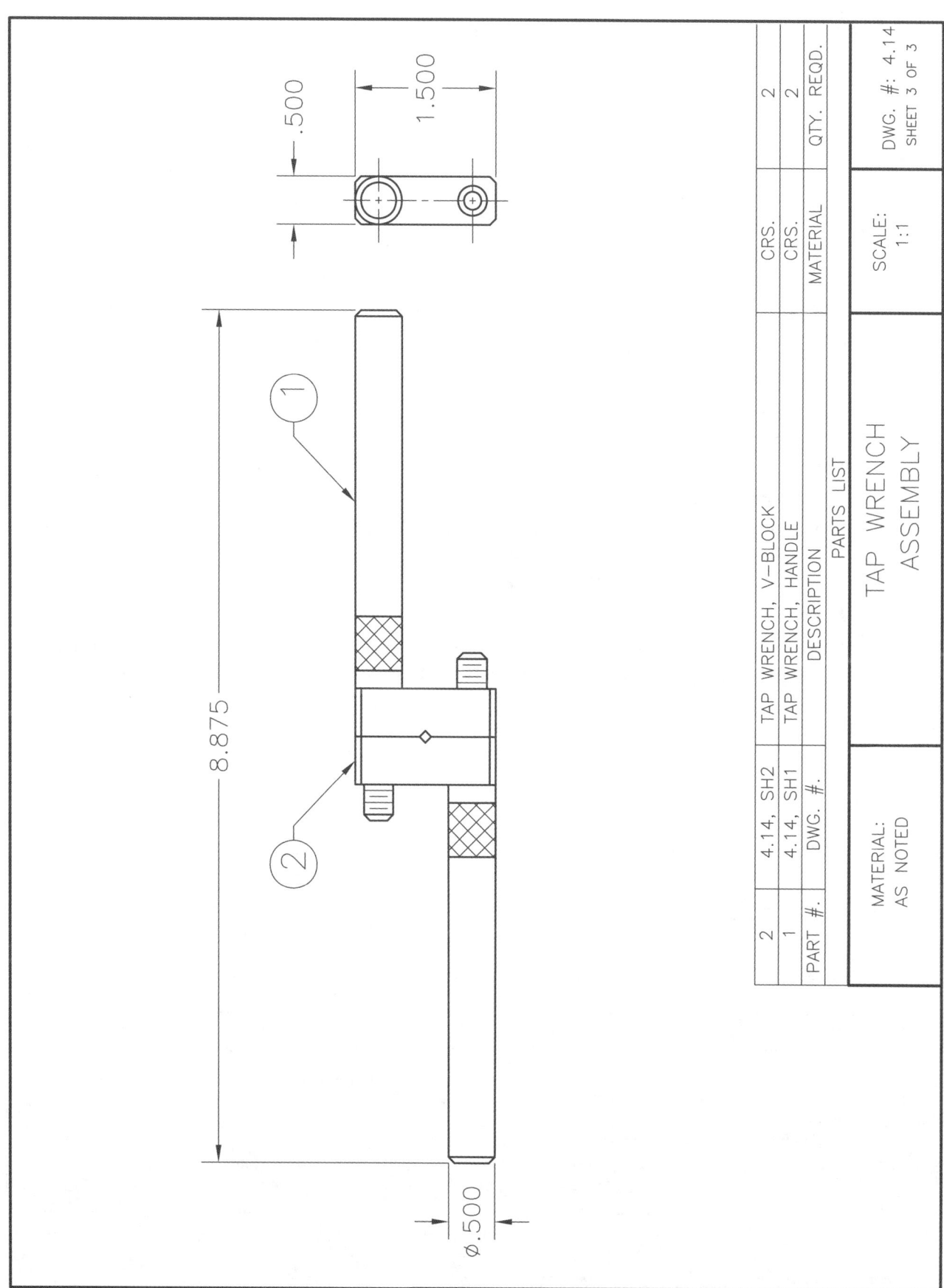

Project 4.15   Step Block                                    Machining Projects

# Project 4.15
# Step Block

Name _____   Date _____

Instructor _____   Period _____

## Order of Operations

- ☐ 1. Measure and cut stock plus facing allowance. Two pieces required.
- ☐ 2. Check trim and alignment of the vertical mill head and vise with a dial indicator.
- ☐ 3. Calculate correct speed and feed for the type and size end mill and material to be used.
- ☐ 4. Mill both blocks to the specified outside dimensions. These blocks are identical.
- ☐ 5. Set up a stock stop and adjust so each step block can be positioned accurately for each cut.
- ☐ 6. Locate the beginning edge of a block with an edge finder and then zero the micrometer collar on the mill table.
- ☐ 7. Mount the end mill, touch off on the top of the block, zero the micrometer collar of the vertical adjustment, then raise the table the required amount.
- ☐ 8. Make the first cut. When this has been accomplished, remove the first block and replace it with the second one. Be sure the second block is seated firmly on the parallels in the mill vise and solidly against the stock stop.
- ☐ 9. Repeat Step 7.
- ☐ 10. Adjust both longitudinal (lengthwise) and vertical for the next step and make the cuts. Repeat this process until both blocks have been cut to their final dimensions.

## Notes

_____
_____
_____
_____
_____
_____
_____
_____
_____

---

**Performance Evaluation—Instructor's Use Only**

Project completed on time?   ☐ Yes   ☐ No
    If No, which steps were not completed _____

Overall performance on project:
    ☐ Excellent   ☐ Good   ☐ Satisfactory   ☐ Unsatisfactory   ☐ Poor

Comments: _____
_____

Instructor's Signature_____

*Machining Projects* **Project 4.15** Step Block

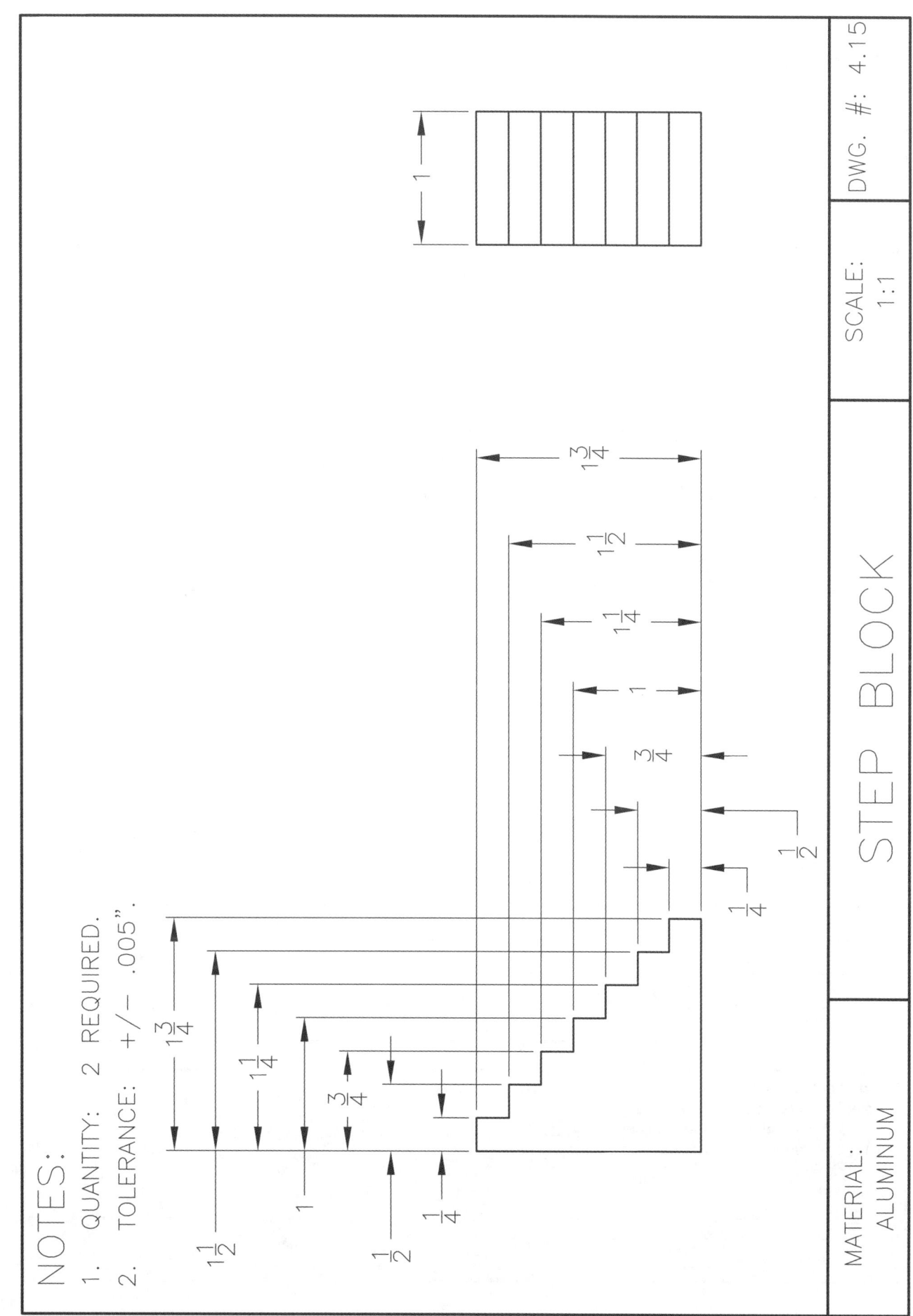

# Project 4.16
# Parallels

Name _____  Date _____
Instructor _____  Period _____

## Order of Operations

Note   The dimensions given in decimal form are the *final* size after grinding. It is recommended that approximately .005 per surface should be allowed for finish grinding after heat treatment.

- ❏ 1. Measure and cut stock plus facing allowance. Two pieces for each size are required.
- ❏ 2. Face to print specifications.
- ❏ 3. Mill side panels to the widths and depths specified for each set.
- ❏ 4. Locate, center drill, drill, and countersink all holes as indicated on the print.
- ❏ 5. Stamp in maker's initials.
- ❏ 6. Heat-treat all pieces according to manufacturer's recommendations.
- ❏ 7. Set up the surface grinder and grind to print specifications. Both pieces of each set should be ground as pairs to ensure uniformity.

## Notes

_____
_____
_____
_____
_____
_____
_____
_____
_____
_____
_____
_____
_____

**Performance Evaluation—Instructor's Use Only**
Project completed on time?   ❏ Yes   ❏ No
  If No, which steps were not completed _____
Overall performance on project:
  ❏ Excellent   ❏ Good   ❏ Satisfactory   ❏ Unsatisfactory   ❏ Poor
Comments: _____
_____

Instructor's Signature _____

Machining Projects    Project 4.16    Parallels

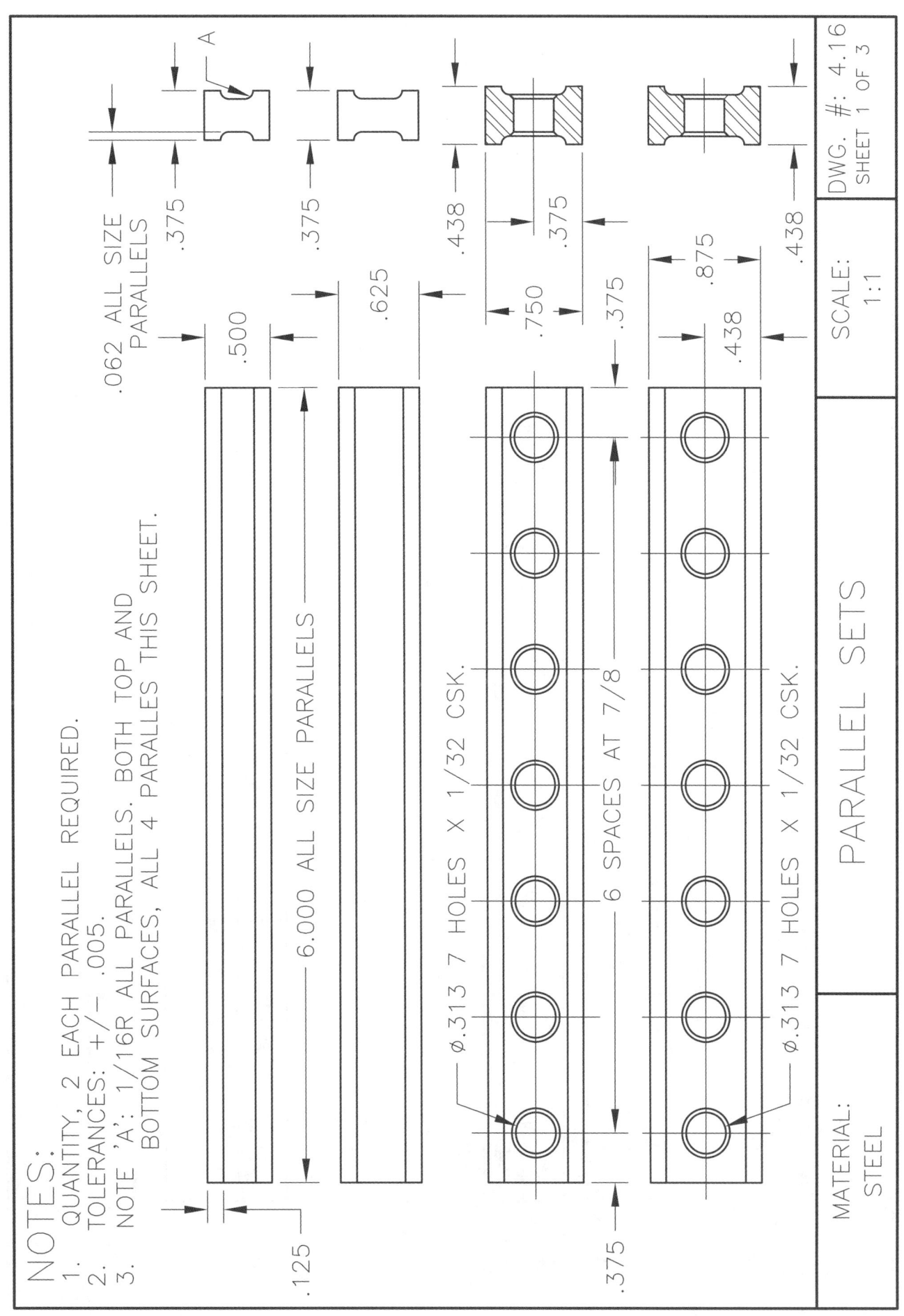

# Project 4.16 Parallels

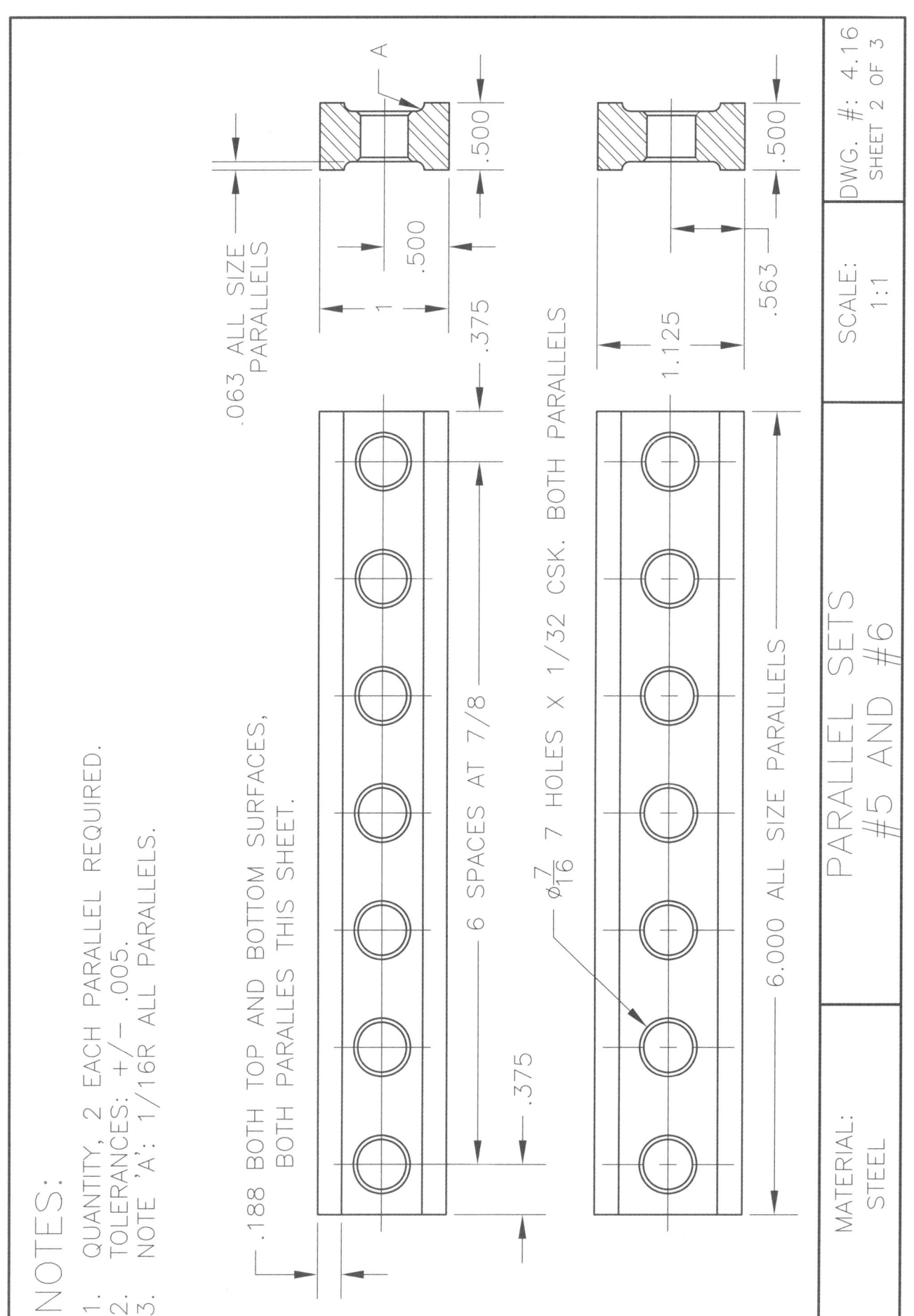

NOTES:
1. QUANTITY, 2 EACH PARALLEL REQUIRED.
2. TOLERANCES: +/- .005.
3. NOTE 'A': 1/16R ALL PARALLELS.

Machining Projects — Project 4.16 Parallels

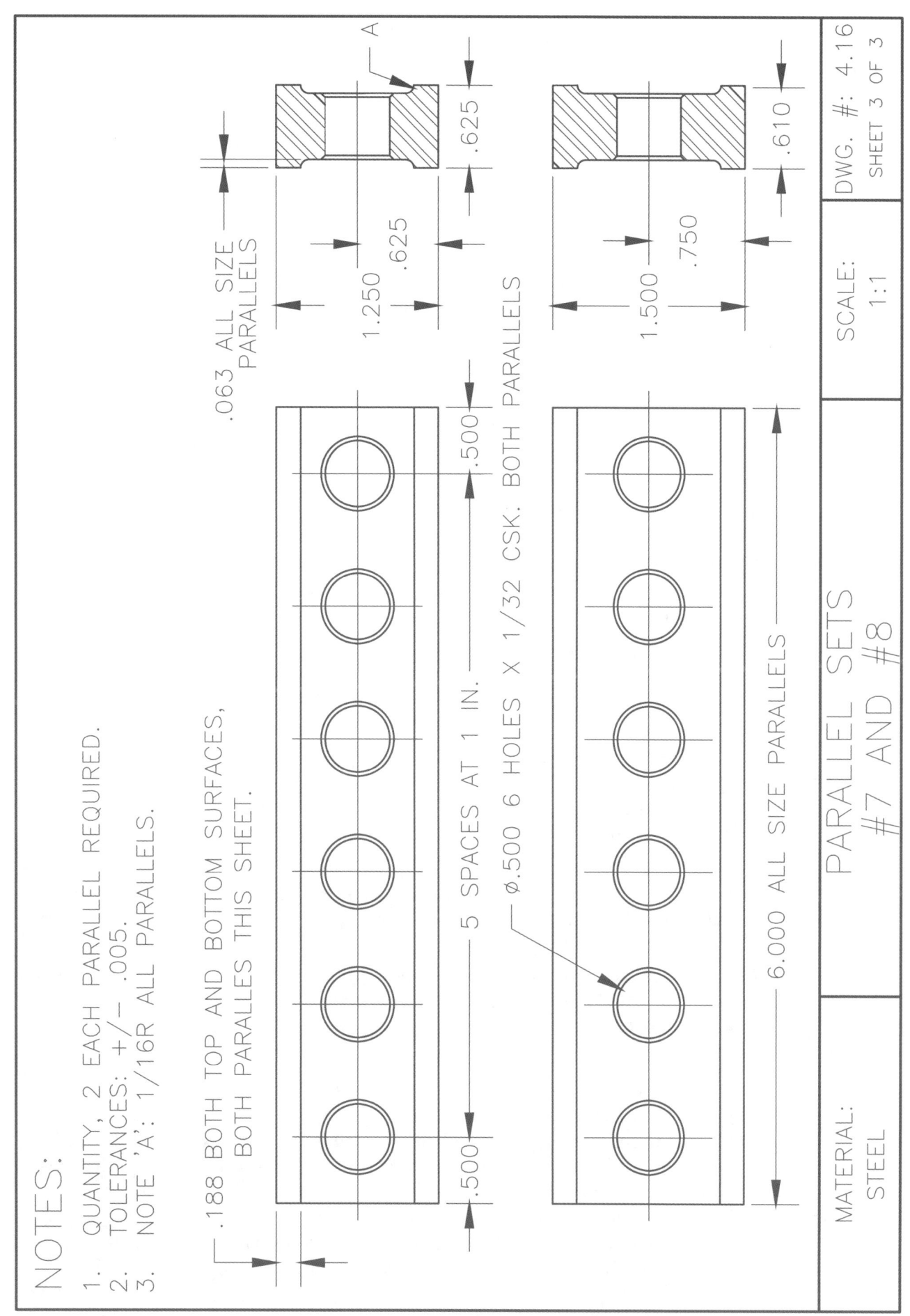

NOTES:
1. QUANTITY, 2 EACH PARALLEL REQUIRED.
2. TOLERANCES: +/- .005.
3. NOTE 'A': 1/16R ALL PARALLELS.

PARALLEL SETS #7 AND #8

MATERIAL: STEEL
SCALE: 1:1
DWG. #: 4.16 SHEET 3 OF 3

Project 4.17   Bolt Welding Jig                                                                    *Machining Projects*

# Project 4.17
# Bolt Welding Jig

Name _____   Date _____

Instructor _____   Period _____

## Order of Operations

❏  1. Measure and cut stock plus facing allowance. Two pieces are required.

❏  2. Check the trim of the mill head and the alignment of the mill vise with a dial indicator to ensure they are within acceptable limits.

❏  3. Calculate the correct cutting speed and feed for the type and size material and end mill to be used.

❏  4. Mount the end mill and secure the workpiece in the mill vise.

❏  5. Face one end of each piece just enough to make them straight, smooth, and square.

❏  6. Deburr the machined ends of the workpieces.

❏  7. Mount a stock stop on the mill table and adjust it so that it will position the workpieces correctly in the vise. The saw-cut end of the pieces must project outward from the vise jaws enough that the ends can be machined to the correct dimension.

❏  8. Place one of the pieces in the vise, push it firmly against the bar of the stock stop, and tighten the vise.

❏  9. Seat the workpiece onto the parallels with a dead blow hammer, then take a light facing cut across the end of the piece. When you have completed the cut, zero the micrometer collars on each end of the mill table—just in case one of the collars is accidentally moved.

❏  10. Measure the length of the piece and record it. Subtract the dimension specified from the current length to determine the amount of material that must be removed.

❏  11. Carefully machine the excess material from the end of the workpiece. When this has been done, remove the first piece and deburr it, then place the second piece in the vise just as was done with the first.

❏  12. With the stock against the stock stop, cut the excess material from the face of the second piece.

❏  13. Turn the workpiece up on edge in preparation for cutting the 1" × 1" slot. It will be necessary to use parallels that will position the piece so that *no less than 1 1/16" of the material extends above the top surface of the vise jaws.*

❏  14. Once the part is seated firmly on the parallels (and butted against the bar of the stock stop), use an edge finder to accurately locate the edge of the piece.

❏  15. Mount a 1" diameter end mill and step off the distance that will place the center of the end mill directly over the center of the slot.

❏  16. With the mill on, raise the table until the end mill contacts the surface of the material.

❏  17. Zero the vertical micrometer collar, set the depth of cut, and make the first cut. The amount of material removed with each cut will be determined by the condition of the mill, the sharpness of the end mill, and the directions of the instructor. It is suggested that cuts of approximately .200 be taken until the desired depth is achieved.

❏  18. When the slot is cut to the specified depth, remove the first block. Now, position and lock into place the second one. It, too, should contact the bar of the stock stop firmly. Repeat the cutting process until the specified depth is reached.

*Machining Projects*        **Project 4.17**     Bolt Welding Jig

- [ ] 19. Deburr both pieces.
- [ ] 20. Rotate the pieces 180°, lay flat and lock and set in the vise, then accurately locate the center of the 1/2" diameter hole.
- [ ] 21. Center drill, drill, and ream the hole as shown. Repeat the process on the second block.
- [ ] 22. Reposition the blocks for cutting the slit.
- [ ] 23. Mount a 1/8" wide slitting saw on an arbor, and cut the slot as indicated on both pieces.
- [ ] 24. Locate the center of the 5/16" hole, drill it through, then counterbore to a depth that will allow a 3/8-16 socket-head cap screw to seat flush with the surface of the block.
- [ ] 25. Ream the remaining part of the top half of the block to 3/8" diameter.
- [ ] 26. Tap the bottom half of the block 3/8-16.
- [ ] 27. Lock and set the pieces on end, locate the center of the holes, and drill and tap the 1/2-20 hole as indicated.
- [ ] 28. Measure and cut 1/2" diameter steel to length plus facing allowance.
- [ ] 29. Face to the correct overall dimension and chamfer both ends.
- [ ] 30. Heat and bend to 90° in the center.
- [ ] 31. Assemble the welding jig.

**Notes**

---

**Performance Evaluation—Instructor's Use Only**

Project completed on time?    ❏ Yes    ❏ No
    If No, which steps were not completed _____
Overall performance on project:
    ❏ Excellent    ❏ Good    ❏ Satisfactory    ❏ Unsatisfactory    ❏ Poor
Comments: _____

Instructor's Signature _____

**Project 4.17** Bolt Welding Jig

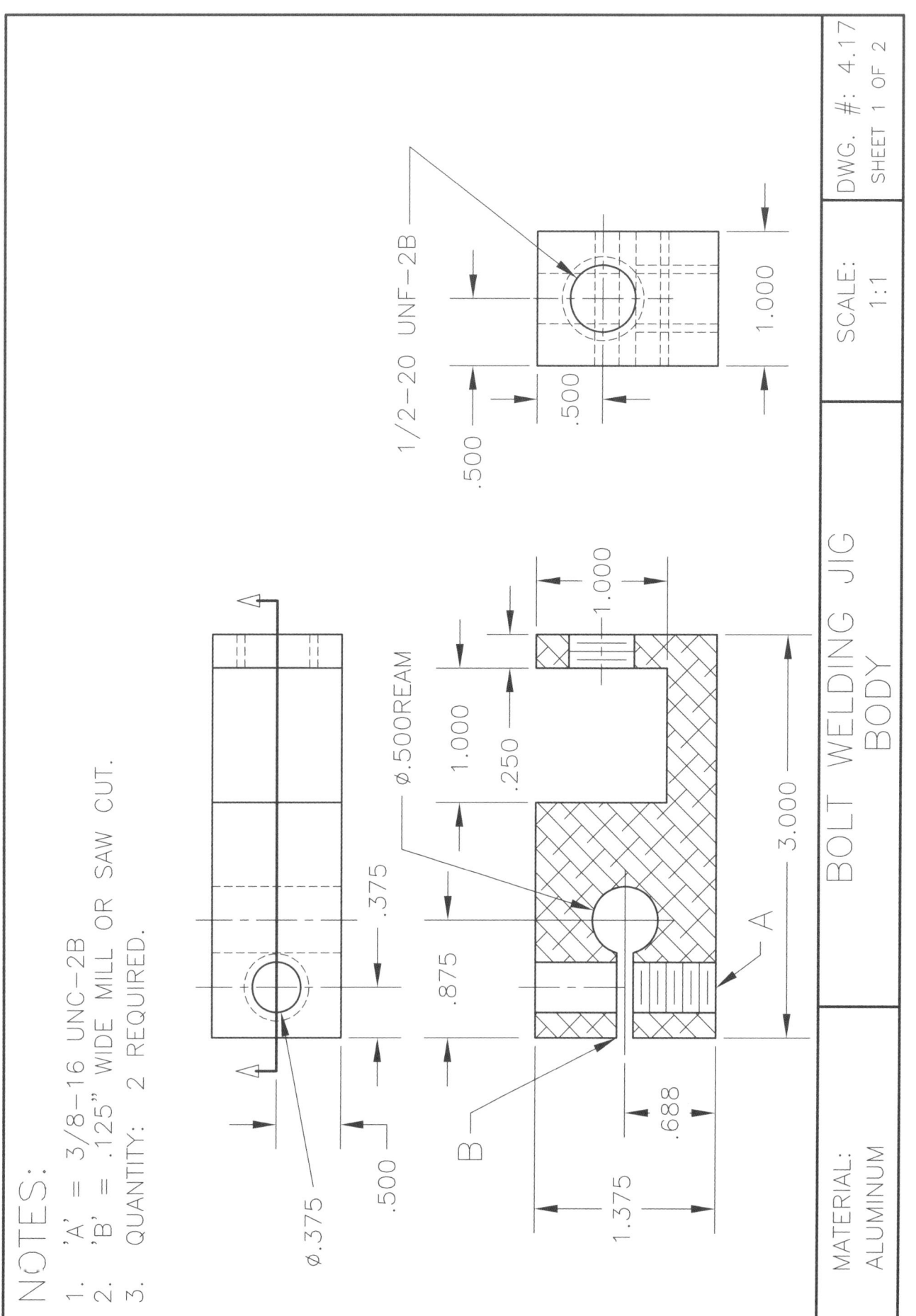

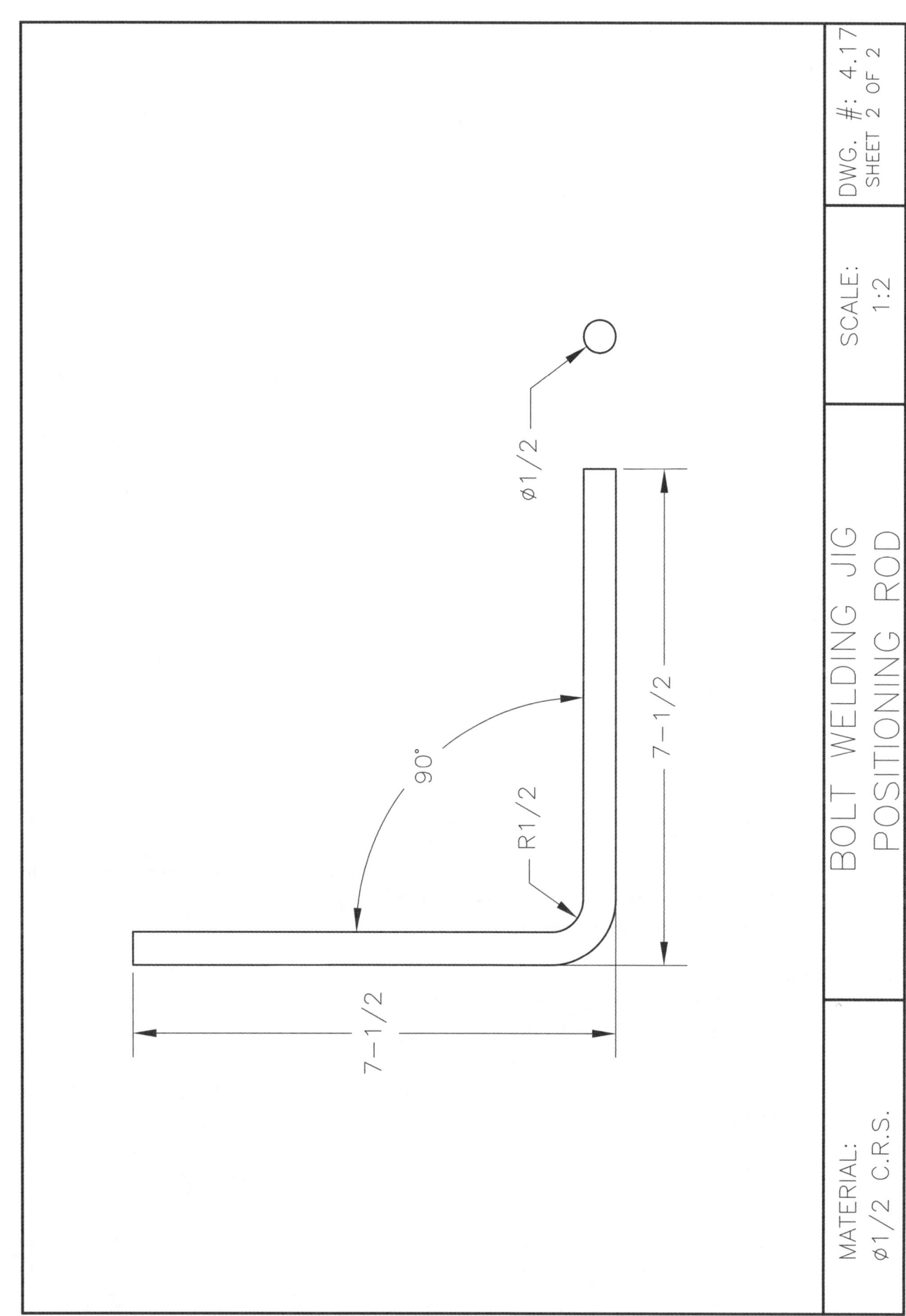

# Project 4.18 Air Engine

| Name | Date |
| Instructor | Period |

## Order of Operations

**Base**

❑ 1. Measure and cut stock plus facing allowance.
❑ 2. Check the alignment and trim of the vertical mill with a dial indicator.
❑ 3. Calculate the correct cutting speed and feed for the size and type end mill to be used and the size and type material to be cut.
❑ 4. Lock the workpiece in the mill vise, set with a deadblow hammer, and face to the specified dimension.
❑ 5. Mount a 1/2" diameter end mill in the spindle and cut the 1/4" radius on all four edges.
❑ 6. Locate the center of the base and drill, countersink, and ream the 3/8" hole.
❑ 7. Deburr all edges.

**Engine Block**

❑ 1. Measure and cut stock plus facing allowance.
❑ 2. Face workpiece to the specified length in vertical mill.
❑ 3. Locate and drill the .375" diameter hole.
❑ 4. Locate, drill, and tap the 3/8-24UNF hole .750" deep.
❑ 5. Locate and drill the .156" diameter hole .375" deep.
❑ 6. Locate and drill the .156" diameter through hole.
❑ 7. Rotate the workpiece 90° in the vise, then locate, drill, and tap the 5/16-18UNC hole .375" deep.
❑ 8. Place the workpiece in the mill vise in a vertical position. Use a precision square to aid in positioning and be sure the bottom of the block is up.
❑ 9. Locate, drill, and tap the 3/8-16UNC hole .625" deep.

**Crankshaft**

❑ 1. Measure and cut stock plus facing allowance.
❑ 2. Face to the specified length.
❑ 3. Turn down one end to .368" diameter × 2.081" length.
❑ 4. Place workpiece in a 4-jaw chuck, offset it .500", and cut to .180" diameter for the length indicated.
❑ 5. Deburr all edges.

Machining Projects · Project 4.18 Air Engine

### Flywheel
- ☐ 1. Measure and cut stock plus facing allowance.
- ☐ 2. Face to the specified length.
- ☐ 3. Turn one end to .870″ diameter × .353″ long.
- ☐ 4. Drill through using a U drill.
- ☐ 5. Place in mill, then locate, drill, and tap the 1/4-20 hole.
- ☐ 6. Deburr all edges.

### Piston
- ☐ 1. Measure and cut stock plus facing allowance.
- ☐ 2. Face to length.
- ☐ 3. Turn to the dimensions shown.
- ☐ 4. Drill #14 drill through as indicated.
- ☐ 5. Deburr all edges.

### Cylinder Block
- ☐ 1. Measure and cut stock plus facing allowance.
- ☐ 2. Face to length in vertical mill.
- ☐ 3. Drill with V drill through where shown.
- ☐ 4. Drill with #19 drill through as indicated.
- ☐ 5. Drill 19/32″ and then ream 5/8″ in end as pictured.
- ☐ 6. Deburr all edges.

### Spring

**Note** A simple mandrel must be made to facilitate the fabrication of the air motor spring. Cut a piece of 5/16″ diameter cold rolled steel 5″ long. Face one end and center drill with a #2 center drill. Approximately 1″ from the other end, drill a 3/64″ diameter hole through the center of the rod.

- ☐ 1. Place mandrel in lathe chuck with 3/64″ diameter hole approximately 1/2″ from the face of the chuck jaws.
- ☐ 2. Engage the point of the live center in the other end and lock the tailstock and tailstock spindle.
- ☐ 3. Place the lathe in neutral.
- ☐ 4. Insert approximately 1/2″ of the end of a piece of .045″ diameter music wire in the 3/64″ hole in the mandrel.
- ☐ 5. Grasp the music wire 8″–10″ for the mandrel and begin turning the spindle by hand.
- ☐ 6. Wrap one full revolution of wire and then begin leading the wire toward the tailstock at a rate that will result in five turns over the space of 1″.
- ☐ 7. When the correct number of turns is achieved, stop movement and complete one full wrap to terminate the spring.
- ☐ 8. Clip the spring material at each end and remove from mandrel.

### Bolt
- ☐ 1. Measure and cut stock plus facing allowance.
- ☐ 2. Face to correct overall length.
- ☐ 3. Turn section indicated to .375″ and thread 3/8-24 for the length shown.
- ☐ 4. Deburr all ends and edges.

**Project 4.18**  Air Engine                                                                 *Machining Projects*

**Inlet Pipe**
- ❏ 1. Measure and cut stock plus facing allowance.
- ❏ 2. Face to the specified dimension.
- ❏ 3. Turn outside to .375″ diameter.
- ❏ 4. Turn one end to .312″ and thread 5/16-18.
- ❏ 5. Drill with #14 drill through.
- ❏ 6. Deburr all edges.

**Finish and Assembly**
- ❏ 1. Sand and polish all parts.
- ❏ 2. Assemble as shown in the assembly drawing.

**Notes**

---

**Performance Evaluation—Instructor's Use Only**
Project completed on time?   ❏ Yes   ❏ No
 If No, which steps were not completed _____
Overall performance on project:
 ❏ Excellent   ❏ Good   ❏ Satisfactory   ❏ Unsatisfactory   ❏ Poor
Comments: _____

Instructor's Signature _____

Machining Projects — Project 4.18 Air Engine

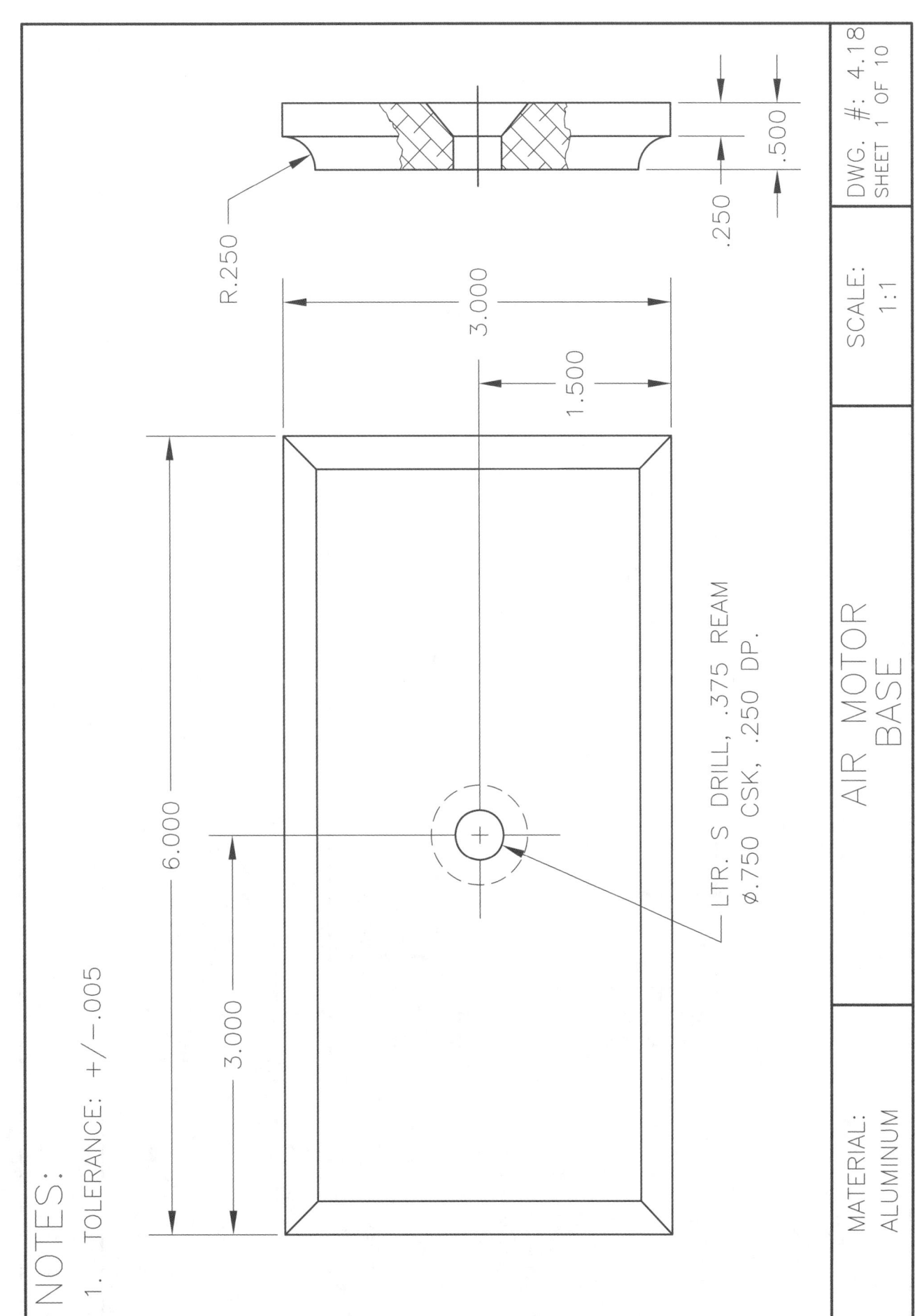

**Project 4.18** Air Engine

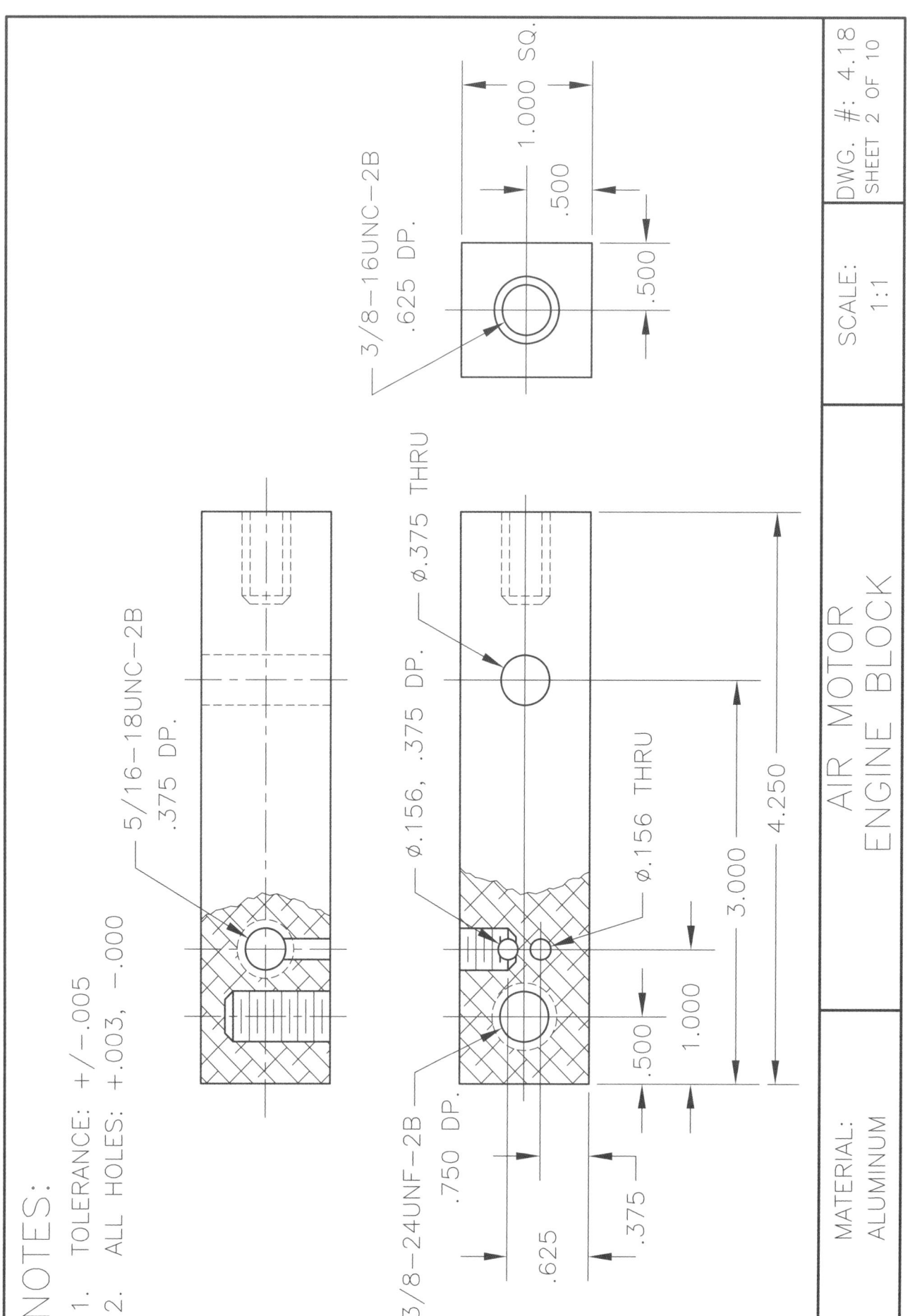

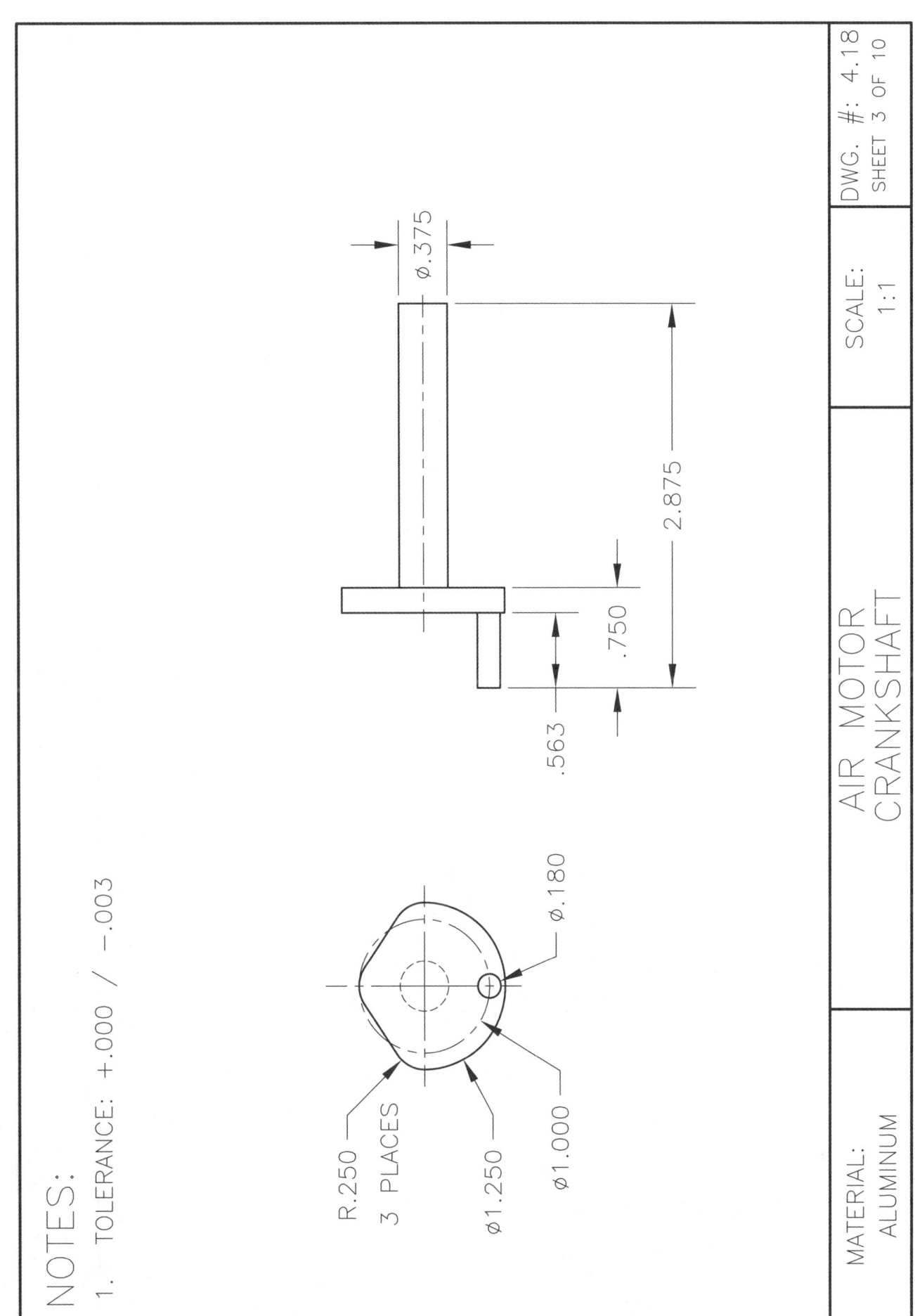

# Project 4.18 Air Engine

*Machining Projects*

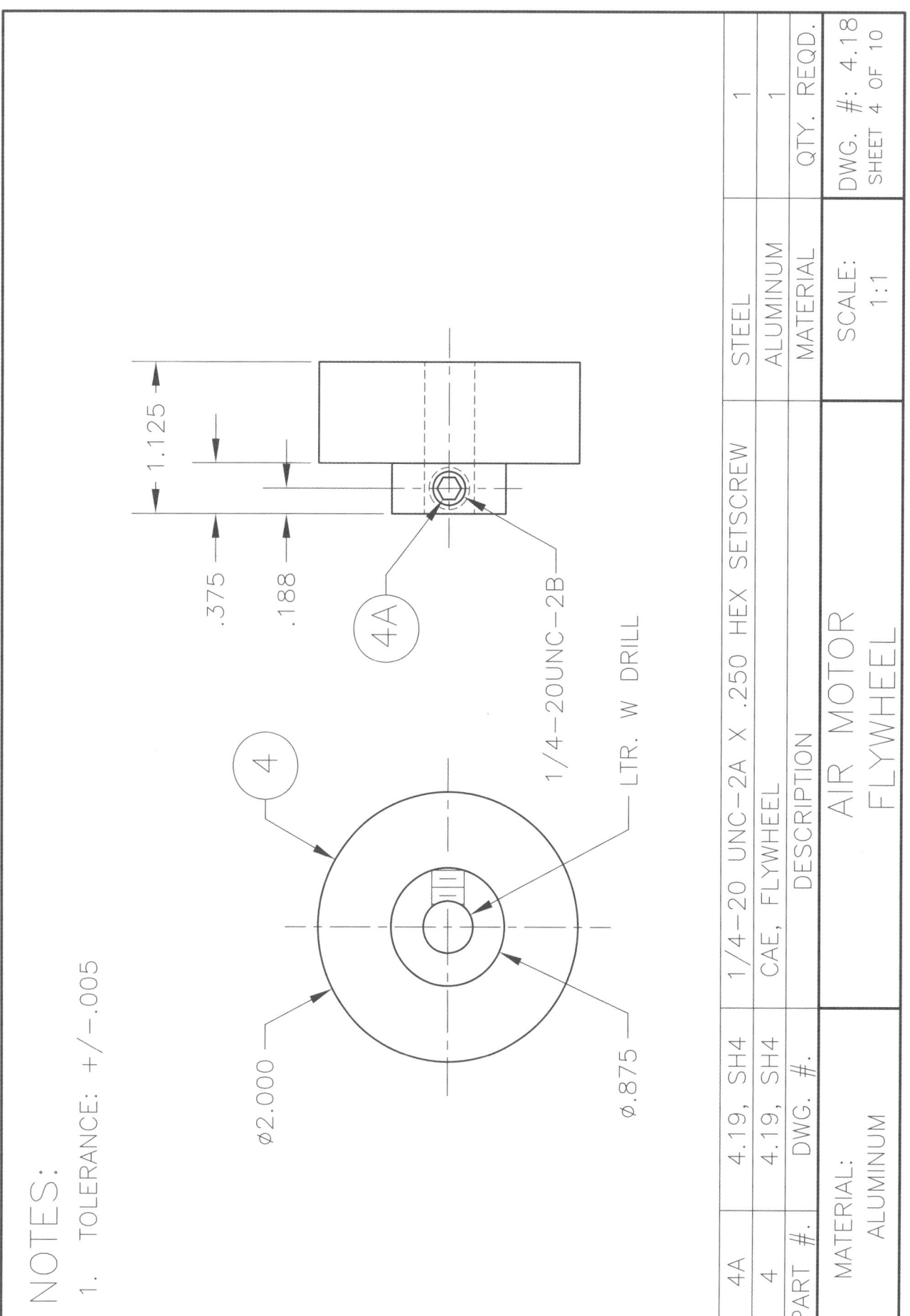

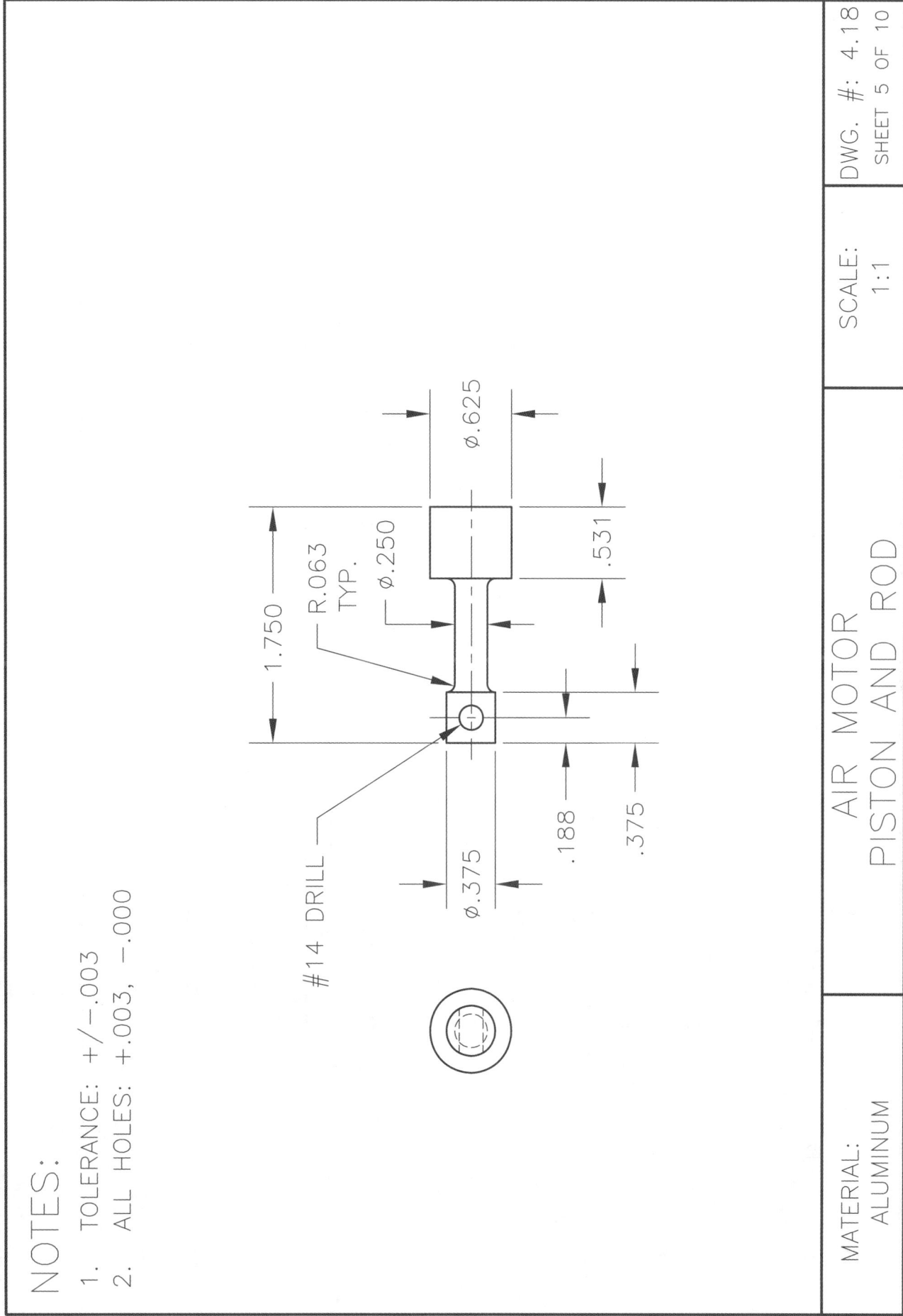

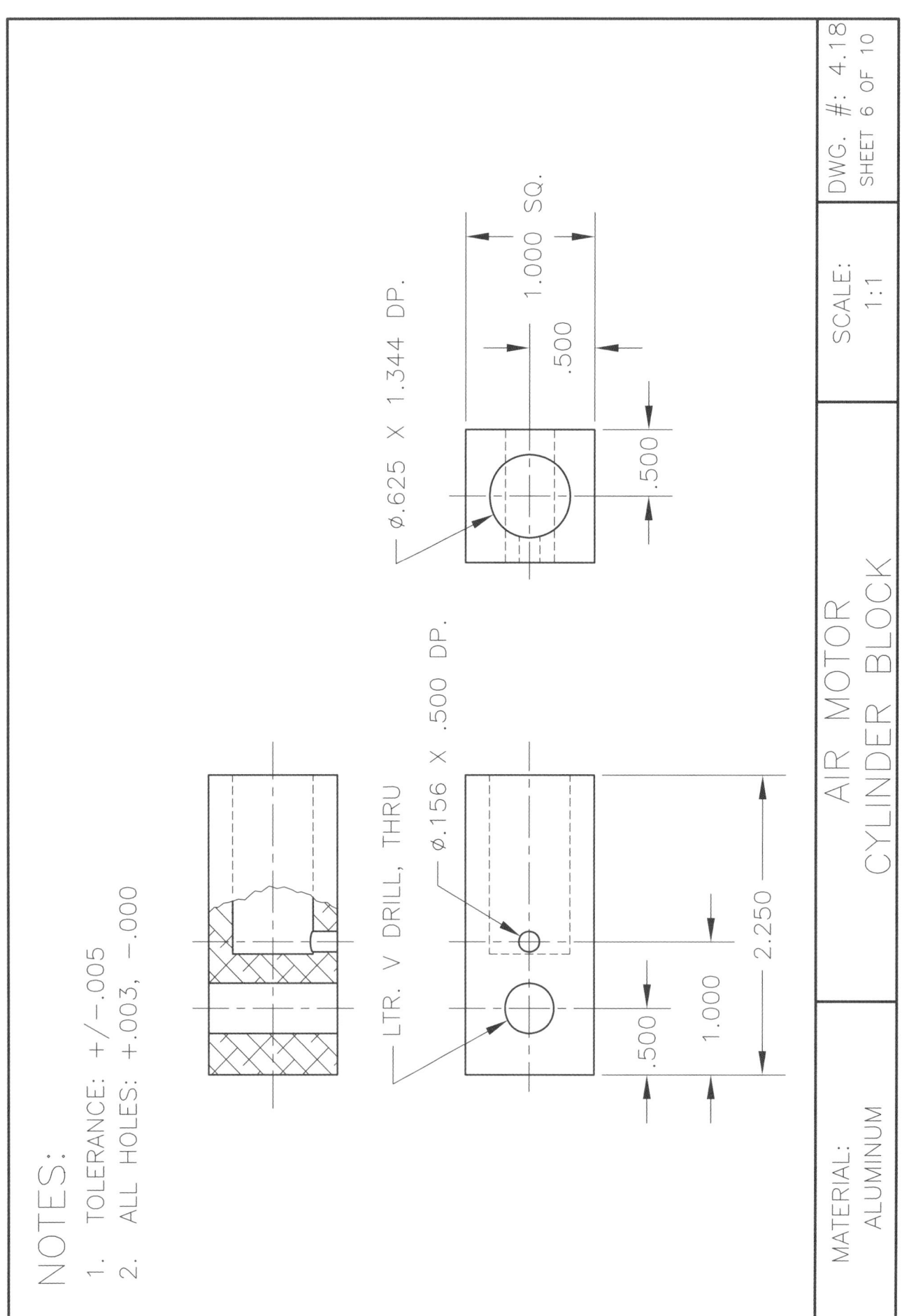

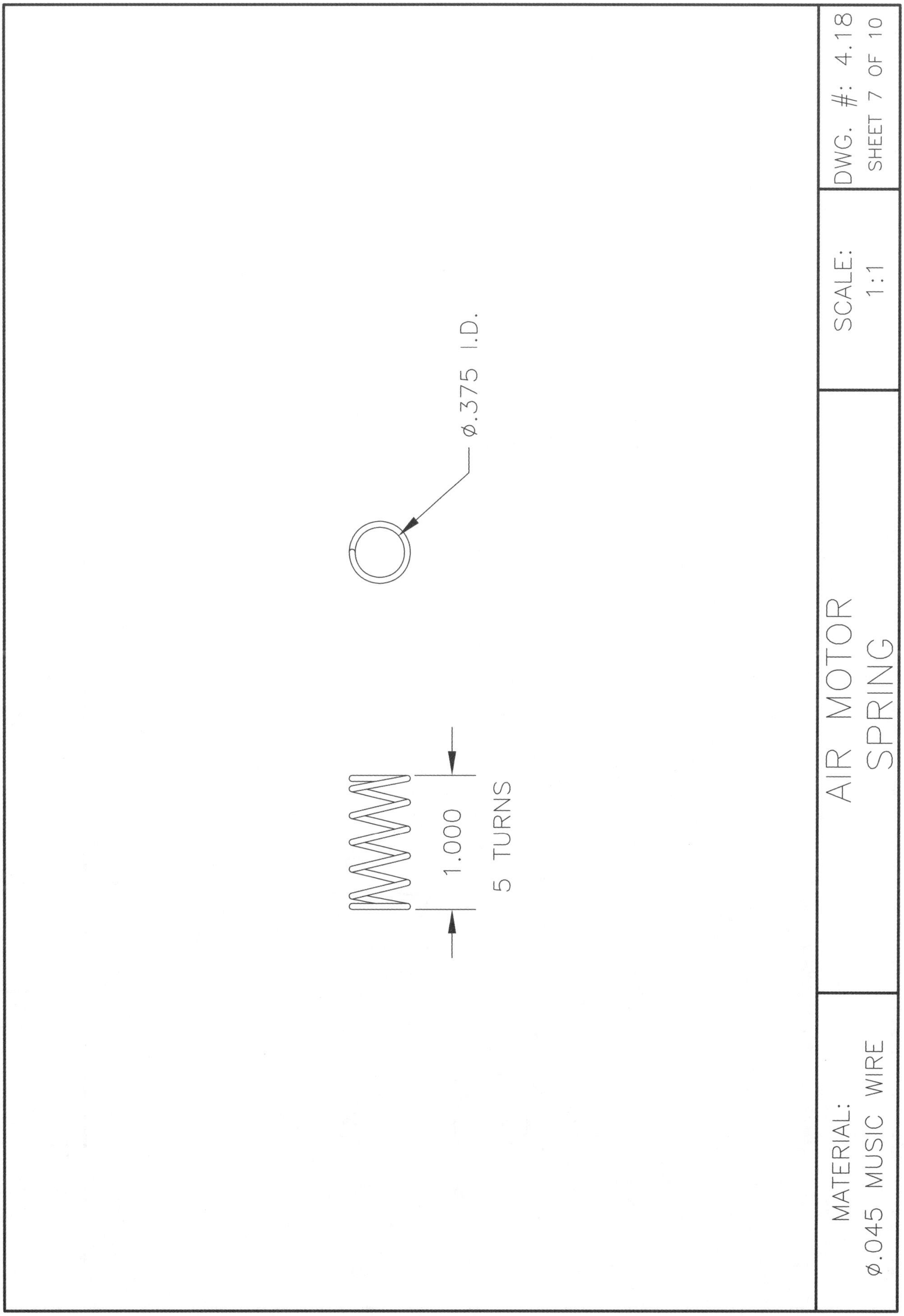

Machining Projects — Project 4.18 Air Engine

Ø.375 I.D.

1.000
5 TURNS

DWG. #: 4.18
SHEET 7 OF 10
SCALE: 1:1
AIR MOTOR SPRING
MATERIAL: Ø.045 MUSIC WIRE

**Project 4.18** Air Engine *Machining Projects*

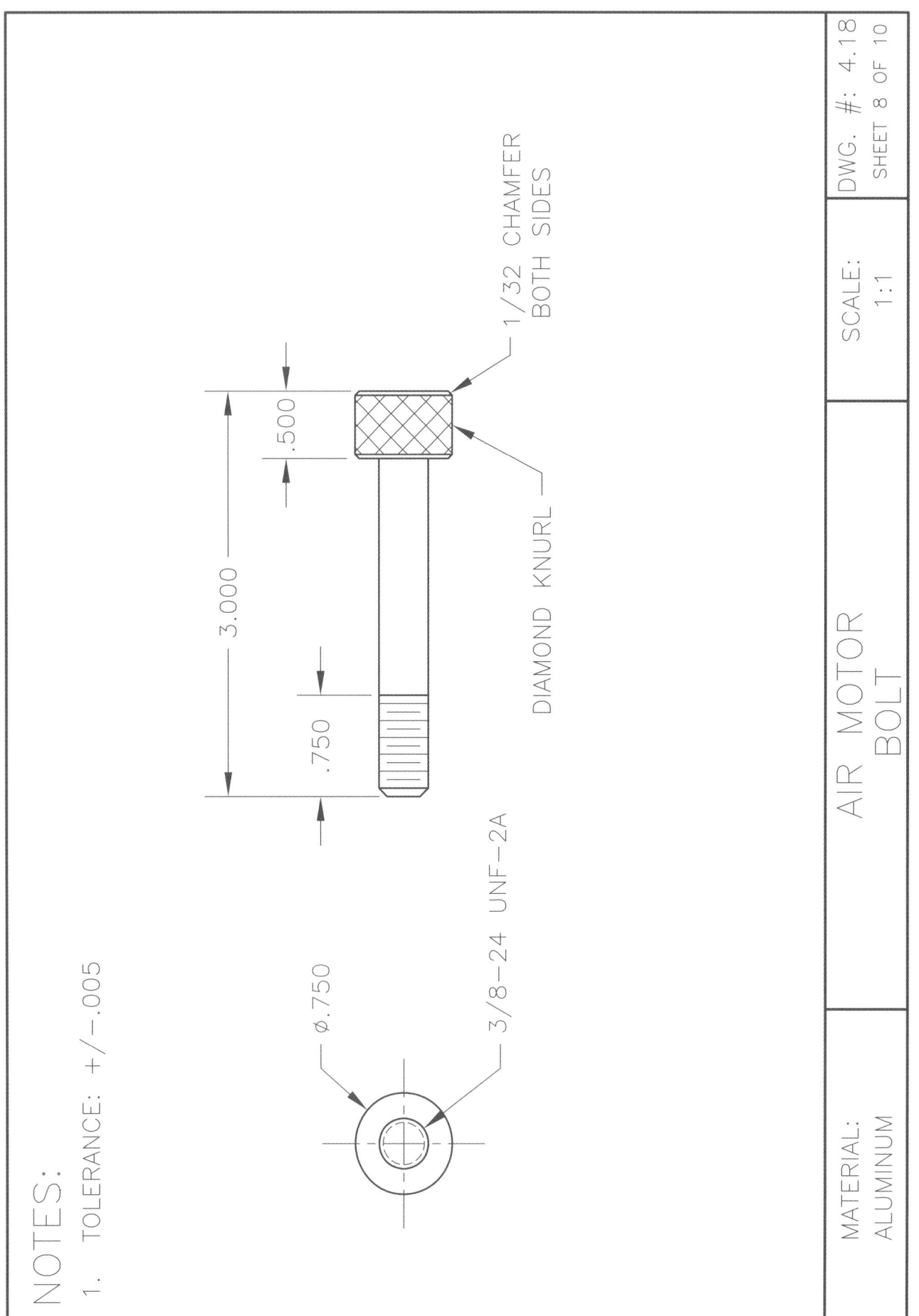

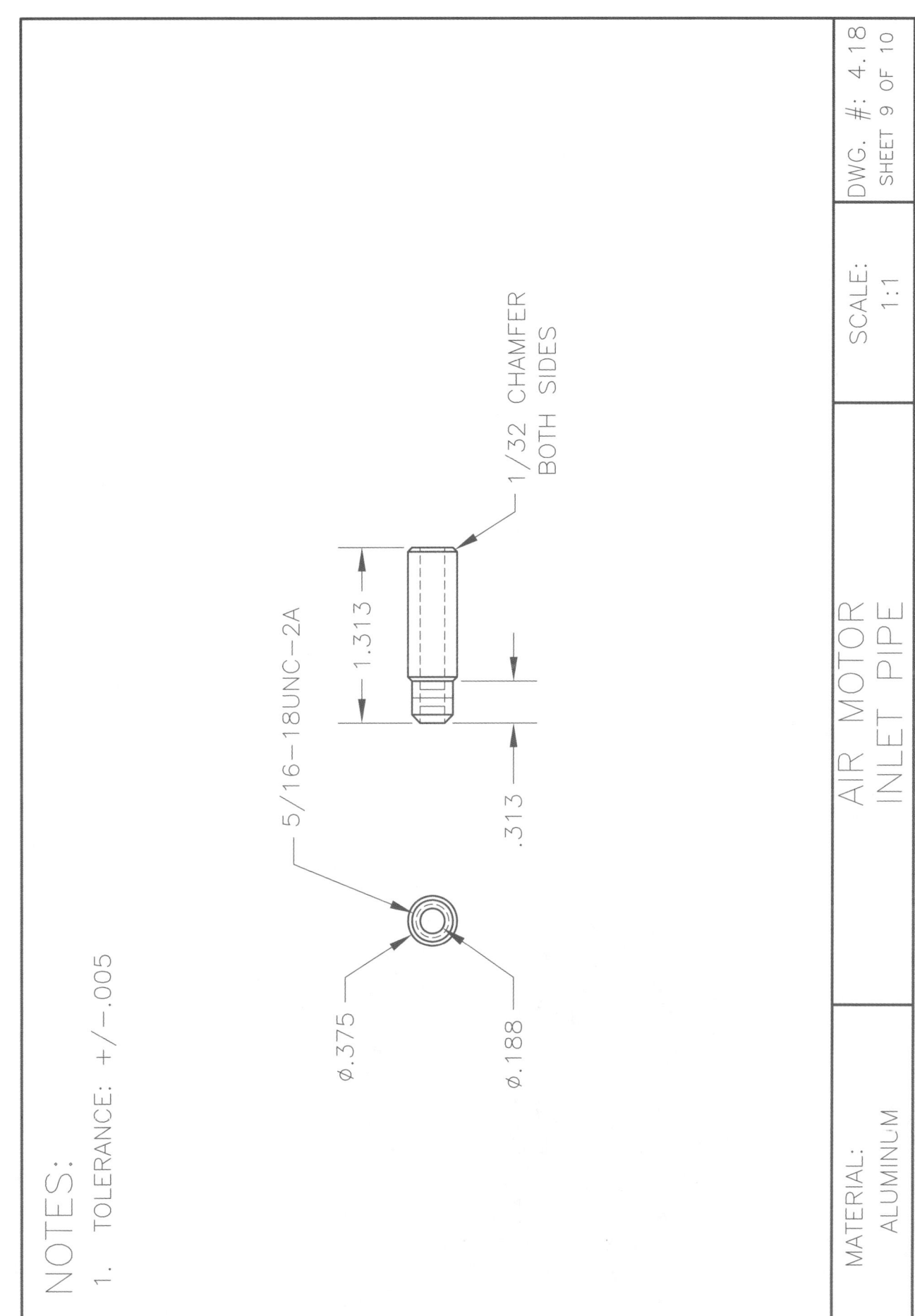

**Project 4.18** Air Engine

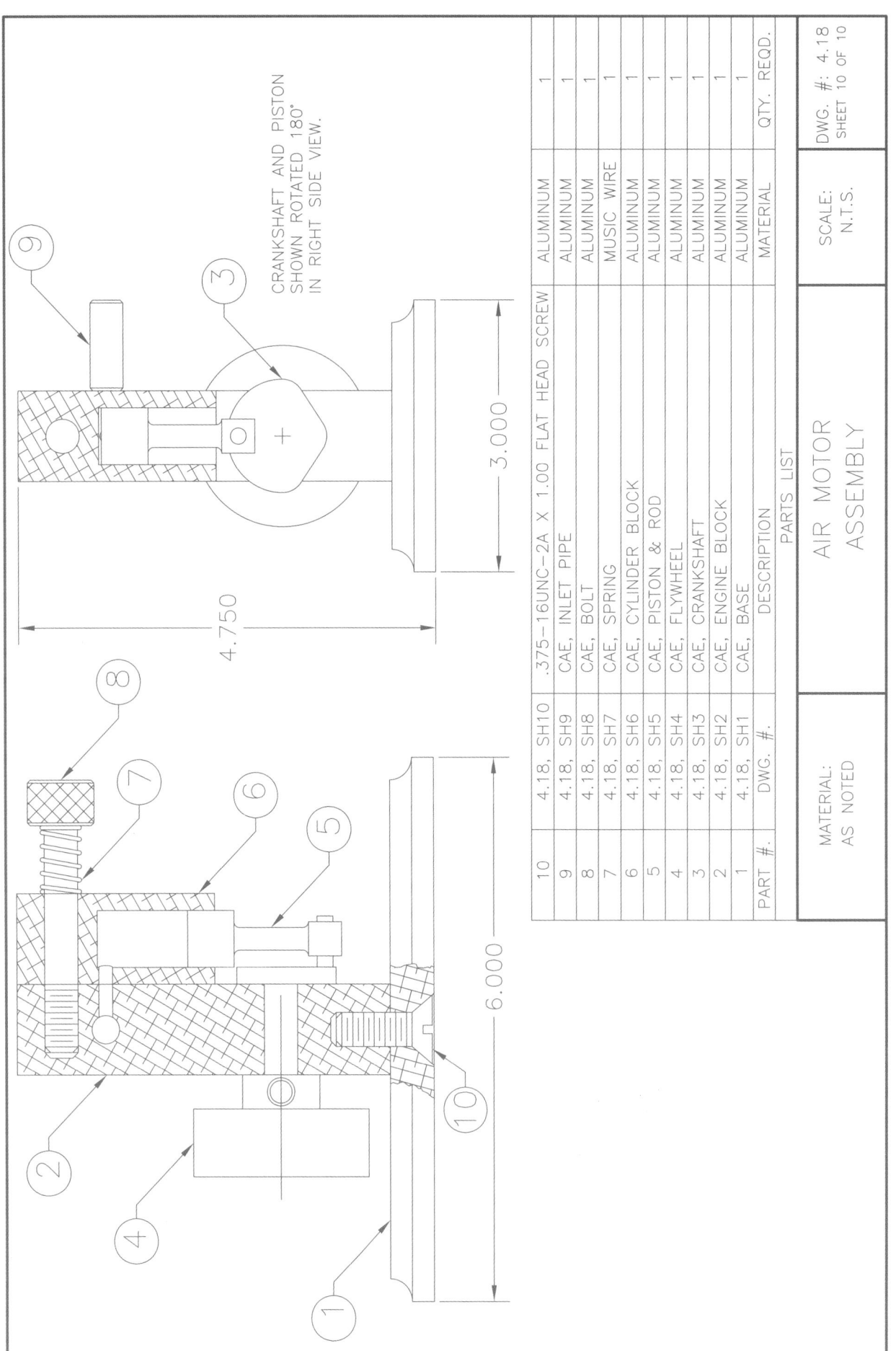